# Mind Maps

# BIOLOGY

First published 2020

The History Press
97 St George's Place
Cheltenham
GL50 3QB
www.thehistorypress.co.uk

British Library Cataloguing in Publication Data.
A catalogue record for this book is available from the British Library.


Design, diagrams and doodles by Lindsey Johns
Project managed by Kate Duffy


ISBN 978 0 7509 9384 5


Printed in China

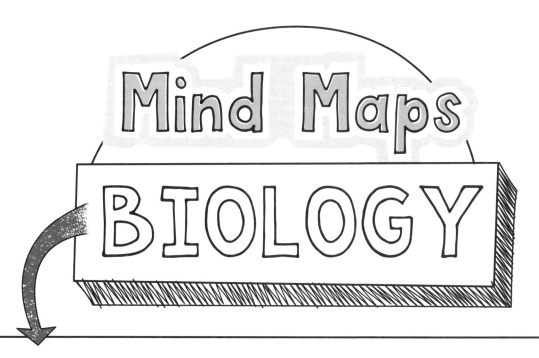

# Mind Maps

## BIOLOGY

## How to Navigate the Living World

DR HELEN PILCHER

The History Press

# Mind Maps

## CONTENTS

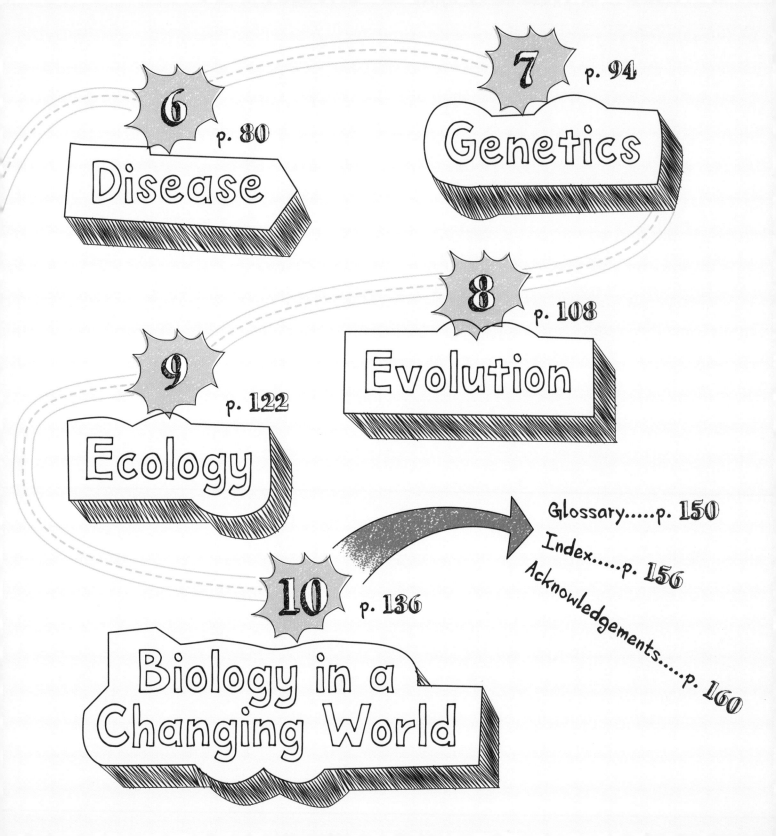

# Introduction

**Mind Maps Biology** is all about the natural world and the living things that it contains. It's aimed at anyone with an interest in life, who wants to know more about biology. It has everything from fungi to ferrets to fossils, with a healthy dash of ecosystems, extremophiles and environmental issues thrown in.

If you're at school or college, it's a useful addition to the classroom and a handy little revision aid. If you're not at school, it's still a great way to engage with one of the most fascinating scientific areas of study. You will learn that the human

brain weighs as much as a chunky guinea pig and that the unique metabolism of the panda means that it can happily eat arsenic-laced bamboo.

This book gives you fascinating facts, surprising snippets and stories from the front line of biology. Did you hear the one about the Hungarian doctor who worried about people not washing their hands? Or the one about the wolves that went to live in an American national park? Both of these biological stories have happy endings.

The book is divided into 10 chapters. Each chapter starts with a mind map, which provides an overview of the contents. The maps are a great place to start when you're exploring a chapter, and a great place to revisit as you make your way through it.

This book takes diverse, complex topics and distils them into smaller, more digestible nuggets. This makes them easier to swallow, easier to understand and easier to remember. Where there is terminology and jargon, it has been 'translated' into simple, accessible

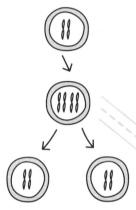

language. Complicated ideas have been simplified to promote clarity and understanding. Whole fields of study have been resolved into visually appealing mind maps that entertain the eyes and stimulate the brain.

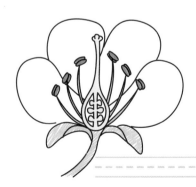

*Mind Maps Biology* is deliberately broad in its scope. It covers many of the topics that are taught at school, as well as some that are not. Biology is an active field of scientific study, so new techniques and discoveries are emerging all the time. With this in mind, *Mind Maps Biology* seeks to balance the well known and established, with the lesser known and novel. So, for example, the classic field of genetics rubs shoulders with the newly formed field of epigenetics, where the environment influences the activity of the genetic code.

This is not a complete or exhaustive guide to the field of biology, rather it includes some of the most interesting areas that are of wide appeal. It's aimed at a broad audience. Any animal with an enquiring brain and an opposable thumb that can flip pages has the potential to be entertained by it.

In its closing pages, this book tackles climate change – one of the most pressing issues of our time. Scientists are certain that our continuing use of fossil fuels is raising the world's temperature and threatening the stability of our planet. Sea levels are rising, the weather is becoming wilder and species are becoming extinct. Change is happening on an epic and global scale, and now scientists think we have entered a new geological era defined by our actions on the planet.

With living things under siege, there has never been a more relevant time to be interested in biology; the study of life. Welcome to *Mind Maps Biology*!

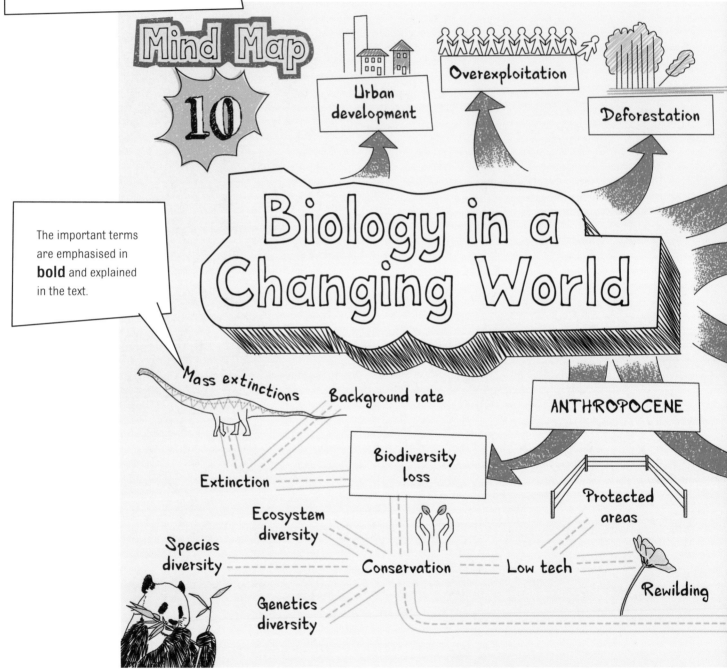

Mind Map

10

Biology in a Changing World

Urban development

Overexploitation

Deforestation

The important terms are emphasised in **bold** and explained in the text.

Mass extinctions

Background rate

ANTHROPOCENE

Extinction

Biodiversity loss

Protected areas

Ecosystem diversity

Species diversity

Conservation

Low tech

Genetics diversity

Rewilding

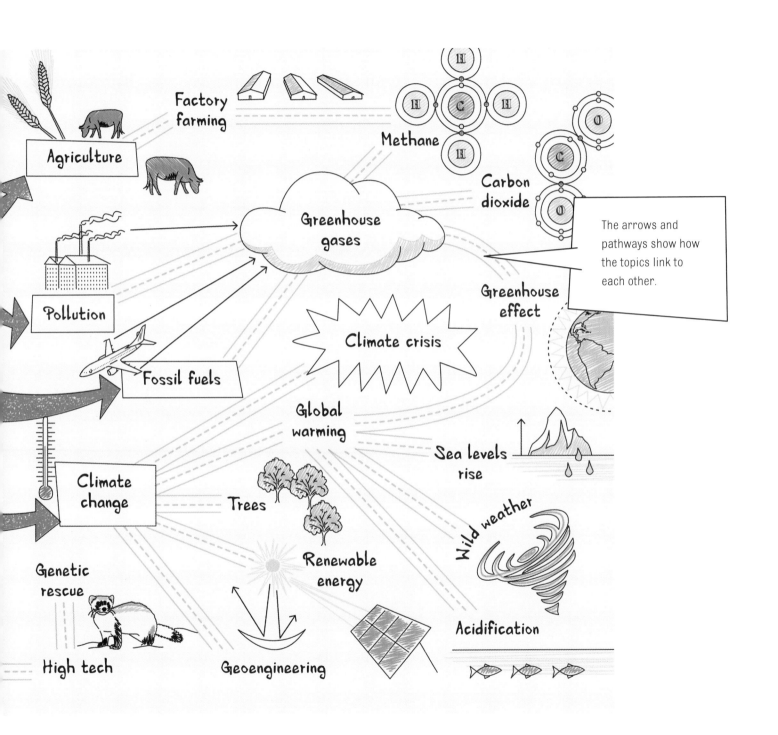

Factory
farming

Agriculture

Methane

Carbon
dioxide

Greenhouse
gases

The arrows and
pathways show how
the topics link to
each other.

Pollution

Greenhouse
effect

Climate crisis

Fossil fuels

Global
warming

Sea levels
rise

Climate
change

Trees

Wild weather

Genetic
rescue

Renewable
energy

High tech

Geoengineering

Acidification

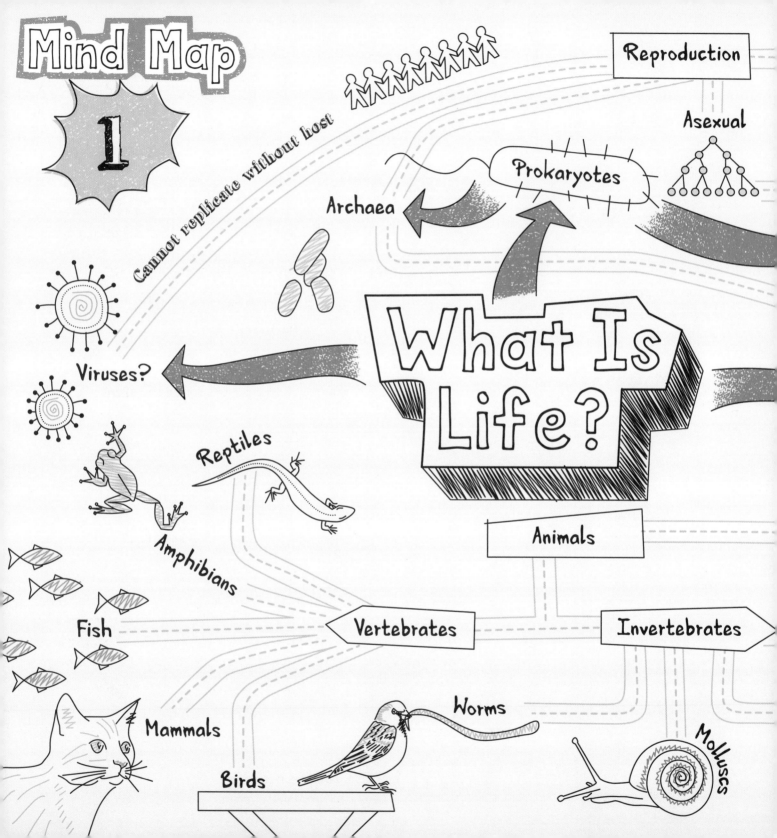

# Mind Map

## 1

Reproduction

Asexual

Cannot replicate without host

Prokaryotes

Archaea

Viruses?

# What Is Life?

Reptiles

Amphibians

Fish

Animals

Vertebrates

Invertebrates

Mammals

Worms

Birds

Molluscs

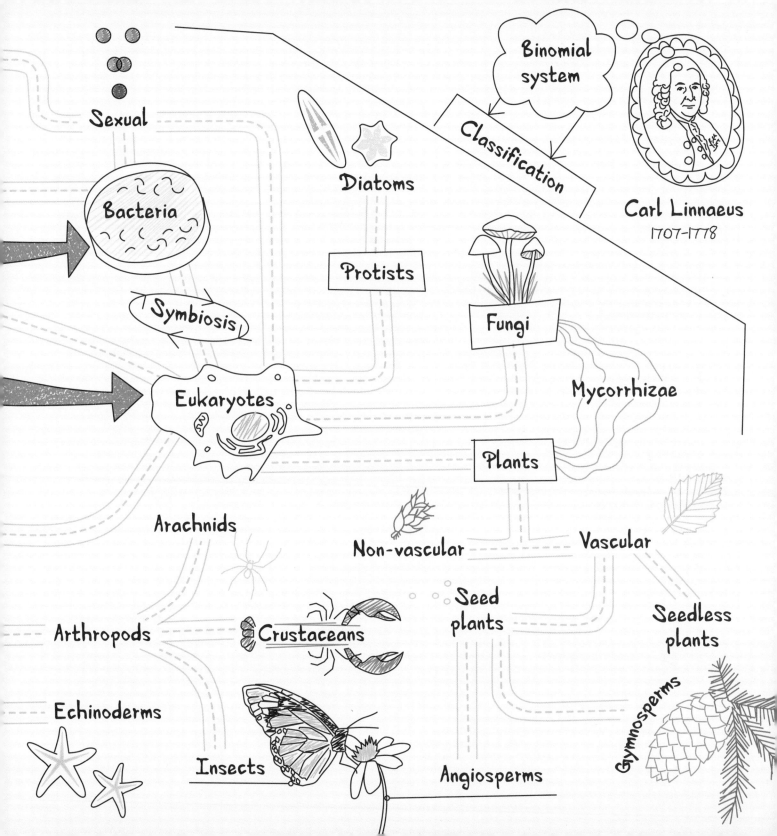

# Living Things

**Homo sapiens**

*Closely related organisms, such as modern humans and Neanderthals, may share a genus name but have separate species names.*

**Homo neanderthalensis**

From fungi to fruit flies, bacteria to butterflies, slime mould to spruce trees, Earth is home to a staggering variety of living things. Biology is the study of life, but it's remarkably hard to define what life is. Instead, biologists recognise life by what living things do.

Living things share a suite of common features. They grow, reproduce, move and respire. They take in and use food, excrete waste products, and can detect and respond to changes in their environment. They are also shaped by the process of evolution.

Although it's intuitively easy to know that a person is alive and a rock is not alive, some things are more difficult to categorise. Viruses, for example, are made from the same building blocks as living things. They feed and reproduce, but can only do so when they hijack the machinery of another living cell. So scientists debate whether they are actually alive or not.

No one knows how many living species there are. Estimates vary, from the millions to the trillions. So far, biologists have identified and named around 1.5 million of these.

Biologists know of about 144,000 species of fungi, 390,000 plant species and 70,000 species of **vertebrates** (animals with backbones). Around 1 million of the world's 5.5 million insect species have been named. There could be trillions of species of microbes, but so far, only a tiny proportion is known. Bacteria, for example, currently make up less than 1% of all described species.

Although many species have familiar, common names, such as jellyfish and sparrow, they also have official, scientific names. Each species has a two-part name. Related species belong to a broader group called a genus, so the first part of the name is the genus and the second part is the species. Humans, for example, belong to the genus *Homo* and the species *sapiens*. Neanderthals are a separate species of human, so they belong to the same genus, *Homo*, but have the species name *neanderthalensis*.

# Classification of Life

A Swedish scientist called **Carl Linnaeus** invented this **binomial system** in the eighteenth century. It is part of a much broader system for classifying life that he also invented. The Linnaean system of **classification** divides life into broad categories, and then subdivides those categories into progressively smaller and more refined groups. It's like identifying a person who lives in a particular apartment, in a building with lots of apartments, in a street with many buildings, in a city with lots of streets.

Today, the original system has been modified. At the top of the hierarchy, all life can be divided into a handful of different domains. These are subdivided into kingdoms, which are split into phyla, which are divided into classes, then orders, then families, genera and species. These different levels are referred to as ranks.

The study of classification is called taxonomy. It is important because it allows scientists to identify, group and properly name organisms. This helps us to understand how living things are related and how they evolved.

In the past, scientists placed organisms in particular categories by comparing obvious features such as structure and function. Today, they also use more sophisticated methods such as genetic analyses. As a result, scientists are continually refining how life is classified.

Earth is around 4.5 billion years old. The first signs of life emerged around 3.8 billion years ago. These were primitive, single-celled organisms, which eventually split to form the three different domains that characterise all life.

The three domains of life are **Bacteria**, **Archaea** and Eukarya. Bacteria and archaea are simple microorganisms, made of single cells. The tiny structures that are found inside these cells are not surrounded by any sort of membrane, so they are known as prokaryotes. Eukaryotes are more complicated organisms that contain membrane-bound structures such as nuclei.

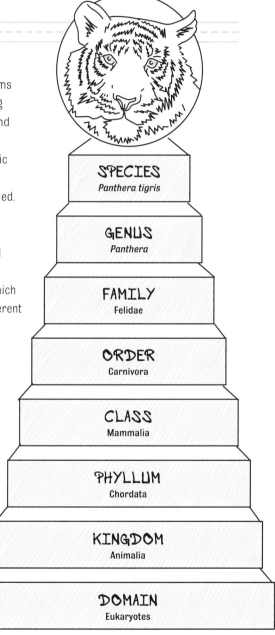

**SPECIES**
Panthera tigris

**GENUS**
Panthera

**FAMILY**
Felidae

**ORDER**
Carnivora

**CLASS**
Mammalia

**PHYLLUM**
Chordata

**KINGDOM**
Animalia

**DOMAIN**
Eukaryotes

Classification of the tiger

# Prokaryotes

For the first 3 billion-or-so years of life on Earth, prokaryotes were the dominant form of life. In many ways, they still are. Scientists estimate that our planet contains around $5 \times 10^{30}$ prokaryotes. That is a staggering 5 followed by 30 zeros. They account for 13% of the world's biomass.

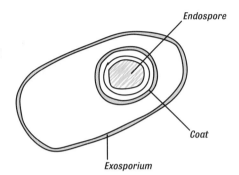

Endospore

Coat

Exosporium

## Bacterial endospore

*The endospore of* Bacillus anthracis, *the bacterium that causes anthrax, is very resilient. It can withstand harsh chemicals, extreme temperatures, and low nutrient levels for decades.*

Bacteria are the most diverse and widespread of prokaryotes. You could fit hundreds of thousands of bacteria onto the full stop at the end of this sentence. They live in diverse environments. They can be found just about anywhere: from water, soil and permafrost to radioactive waste. They have been found deep down in the Earth's crust and high up in the stratosphere. They also live in and on animals and plants, where they can be harmful and helpful.

When conditions are harsh, some bacteria form resilient structures called endospores. Some endospores can survive disinfectants, boiling and extreme freezing. They enable the bacterium to lie dormant for long periods of time, then 'reawaken' when conditions improve. Scientists say they have germinated endospores found in the tombs of Egyptian pharaohs and from bees trapped in 30-million-year-old amber.

Archaea tend to be smaller than bacteria. They share some features with bacteria, and others with eukaryotes. They also have unique traits of their own. For example, they have evolved special metabolic capabilities that enable them to use various energy sources, including organic compounds like sugars, inorganic substances like metal ions and hydrogen gas, and the Sun's light rays.

Compared with bacteria, which are relatively easy to grow and study in the lab, little is known about archaea. Archaea have been found living in volcanic springs, boiling deep-sea hydrothermal vents and the salty Dead Sea of Israel. When they were first defined over 40 years ago, scientists thought they were extremophiles, but since then, they have been found almost everywhere.

## The most common shapes of bacteria

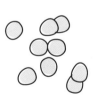

### Spherical
*e.g.* Cocci *bacteria*

### Rod-shaped
*e.g.* Bacilli *bacteria*

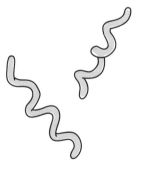

### Spiral
*e.g.* Spirillum *bacteria*

## THE CRUCIAL ROLE OF PROKARYOTES

Prokaryotes play a vital role in Earth's biosphere. If humans disappeared overnight, life would continue, but if all the prokaryotes disappeared, life would be unable to continue. We often think of prokaryotes as a nuisance, because some cause disease, but many more are either harmless or beneficial to eukaryotic life forms.

They help other organisms by acting as recyclers. Some prokaryotes are decomposers. As they break down waste products and dead organisms, they unlock supplies of carbon, nitrogen and other elements. They also convert molecules into forms that can be used by other organisms. For example, cyanobacteria – which are also known as blue-green algae – convert carbon dioxide into simple sugars, which can then be passed up the food chain. Cyanobacteria are photosynthetic (see page 62), so they also produce oxygen, and some prokaryotes convert atmospheric nitrogen into a form that other organisms can use.

Prokaryotes also have important ecological functions. Deep in the ocean, for example, hydrothermal vents support a surprising array of life, including **worms**, snails, shrimps and **fish**. Bacteria are the base of this food chain. They harvest chemical energy from the metals, gases and minerals that spew from the vents, and when they are consumed, they pass on this energy.

Often, they live in close contact with other, much larger organisms. This biological interaction is called **symbiosis**. Symbiotic relationships can be mutualistic (both species benefit), commensal (one species benefits but the other is unaffected) or parasitic (one species is harmed).

The eyelight fish has special organs under its eyes that are filled with bioluminescent bacteria. This relationship is mutual, because the bacteria receive nutrients from the fish, and the fish uses the light to attract prey and signal to mates. Meanwhile, in our guts, 'good' bacteria can help to protect us from certain diseases, and boost the functioning of our immune systems.

## The eyelight fish

*Bioluminescent bacteria act like a head torch, attracting prey and mates. The bacteria receive nutrients from the fish.*

# Eukaryotes

Eukaryotes are the most diverse domain of life. This domain contains many of the multicellular life forms that we are familiar with, from mushrooms to monkeys to maple trees, but it also contains many simple, single-celled organisms, such as phytoplankton. These single-celled eukaryotes vastly outnumber their multicellular counterparts, but because the multicellular organisms are bigger, their collective bulk accounts for more than 80% of the world's biomass.

The eukaryotes are divided into four different kingdoms: protists, fungi, plants and animals. Eukaryotes are different from prokaryotes because their cells contain membrane-bound structures, such as nuclei, mitochondria and chloroplasts (see page 29). Multicellular eukaryotes, such as worms and whales, are also made up of many different types of cells and tissues.

The emergence of multicellular life is a milestone in the story of life on Earth, yet biologists struggle to work out when and where it happened. At some point, single cells started to group together and acquire specialised functions. This gave them new skills, for example increased mobility, so they could avoid predators and find food more easily.

Complex eukaryotic life, with structured bodies and multiple cell types, is clearly visible in the fossil record from around 600 million years ago. The first eukaryotes must have evolved before then, but because their small, simple bodies degraded easily, they are not easily found in the fossil record.

In 2017, researchers found simple, fungus-like fossils in rocks from South Africa that are 2.4 billion years old. When they were alive, these organisms would have lived on the sea floor, so the find suggests that eukaryotes arose not on land, but in the sea. These could be some of the world's earliest eukaryotes.

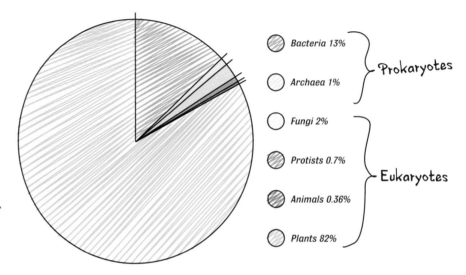

Bacteria 13%

Archaea 1%

} Prokaryotes

Fungi 2%

Protists 0.7%

Animals 0.36%

Plants 82%

} Eukaryotes

## Breakdown of the world's biomass

*Although we like to think we're mighty, animals make up just a tiny proportion of the world's biomass. Plants make up the majority.*

## Fungus-like forms

*Fungus-like forms have been found in rocks that are 2.4 billion years old. These sea-dwelling creatures could be some of the earliest eukaryotes.*

## PROTISTS

In 1674, Dutch scientist Antonie van Leeuwenhoek peered down his homemade microscope and saw a myriad of tiny, single-celled creatures, which we now call **protists**.

Protists come in many shapes and sizes. Some look like mini trumpets or jewels. Some have tiny hair-like structures, called cilia, that propel them along. Others have whip-like tails called flagella. Most are single-celled, but some species live as colonies or are multicellular. Protists are more diverse — functionally and morphologically — than fungi, animals and plants.

### Protists of different shapes

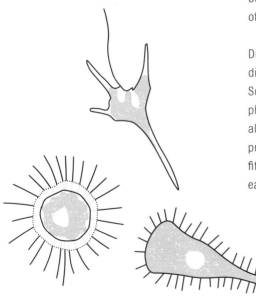

Van Leeuwenhoek said, 'No more pleasant sight has met my eye than this, of so many thousands of living creatures in one small drop of water'.

Protists obtain their nutrition in many different ways. Some make their own food from inorganic substances, using light as their energy source. Some are heterotrophs: they absorb organic molecules or ingest larger food particles. Others, which combine the two approaches, are called mixotrophs.

**Diatoms** are tiny organisms that mostly live in watery environments such as oceans, lakes, ponds and streams. Diatoms are a type of protist with glassy silica shells. Dubbed 'the jewels of the sea', they are one of the most abundant organisms living in the ocean. One bucketful of seawater contains millions of these microscopic algae.

Diatoms affect global levels of carbon dioxide and oxygen. They convert the Sun's energy into chemical energy via photosynthesis. As a result, they produce almost half the organic compounds produced in the ocean, and generate a fifth of the oxygen produced on Earth each year.

### Diatoms: the jewels of the sea

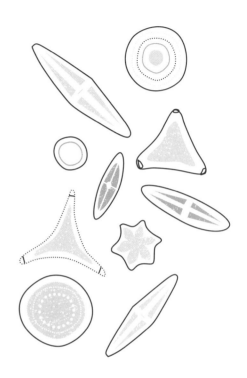

They provide food for other protists and aquatic creatures, and those that are not eaten take decades or more to decompose. Their bodies sink to the ocean floor where they accumulate in dense layers, locking away the carbon they contain for long periods of time.

With atmospheric carbon dioxide ($CO_2$) levels rising, scientists are now considering fertilising the ocean with diatom-nourishing chemicals, to promote this carbon-sequestering ability.

# Fungi

Yeast, moulds and mushrooms are all members of the fungi kingdom. Some fungi are single-celled, but most have complex, multicellular bodies. They can't make their own food like plants and algae, and they can't consume food like animals. Instead, they are heterotrophs.

Some fungi obtain their nutrients by secreting enzymes, which break down complex molecules into smaller organic compounds that can be absorbed more easily. The enzymes are varied, meaning that fungi can digest compounds from a range of sources. Parasitic fungi, like the ones that cause athlete's foot, absorb nutrients from the cells of living hosts. Fungi that act as decomposers absorb nutrients from non-living organic material like leaves and corpses.

Fungi can be found on land and in the water. They are essential for the well-being of most ecosystems. They break down organic matter and recycle nutrients, making them accessible to other organisms.

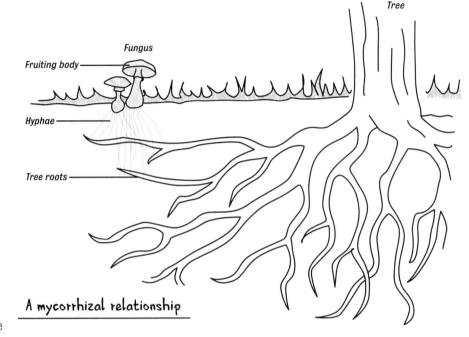

## A mycorrhizal relationship

*Fungal hyphae make contact with tree roots. This helps them communicate and swap nutrients.*

Fungi aren't mobile, so they spread by producing spores that they disperse from their fruiting bodies. The fruiting bodies are the bits of a mushroom that you can see above ground. They are packed full of tiny filaments called hyphae, which extend into the ground to create a mass of interwoven threads. This network of filaments is known as the mycelium.

As the underground filaments grow, they absorb nutrients. The filaments of **mycorrhizae** fungi form intimate associations with tree roots. The fungi supply their hosts with water and essential nutrients, while the trees provide the fungi with the sugars they need for growth. This is an example of a mutualistic relationship.

Most land-living plants have mycorrhizal relationships, which connect the plants to one another underground. Fungi can even transfer sugars between trees of different species, so the nutrients from one tree can nourish the cells of another. In addition, it's thought that plants use mycorrhizae to communicate, so they can, for example, warn each other of an impending insect attack.

# Plants

The plant kingdom is bountiful and diverse. It includes flowering plants, conifers, ferns, hornworts, liverworts, mosses and green algae.

Plants are the basis of most of Earth's ecosystems. They make up more than 80% of the world's biomass, and are a source of food, habitat and products such as timber and medicine.

Green plants make their own food via photosynthesis (see page 62). Along the way, they lock up carbon and release oxygen into the atmosphere. Land-living and marine plants produce much of the oxygen we breathe.

Plants evolved from single-celled green algae over 500 million years ago. The first plants lived in water. The first land plants appear in the fossil record around 470 million years ago. After that, they diverged to form several major groups.

Vascular plants, like ferns and trees, have an extensive system of vascular tissue. This is made of cells that are joined together to form tubes that transport water and nutrients through

## Vascular plants

*Ferns*

*Vascular plants contain tube-like vessels that they use to transport water and nutrients*

the plant. Most of the plants that are alive today are vascular. Plants that do not have a defined transport system, such as liverworts and hornworts, are called **non-vascular** plants.

In the early stages of plant evolution, non-vascular plants dominated the landscape until **vascular** plants emerged around 425 million years ago. These early vascular plants lacked seeds, but they did have well-developed vascular systems. As a result, they became the first plants to grow tall.

Today's vascular plants can be split into two groups. There are **seedless plants** like ferns and horsetails, which spread via windblown spores, and there are **seed plants**.

## Non-vascular plants

*Mosses*

*Non-vascular plants have no vascular system*

*Liverworts*

A seed is an embryonic plant enclosed in a protective outer coat, usually with a supply of food. Seed plants are either **gymnosperms**, like cycads and conifers, or angiosperms, like grasses, shrubs and most trees. The seeds of gymnosperms are not enclosed. The seeds of **angiosperms** develop inside specialised chambers within the flowers.

## Anatomy of a seed

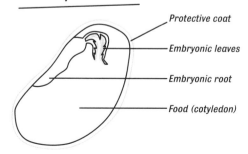

- Protective coat
- Embryonic leaves
- Embryonic root
- Food (cotyledon)

# Animals

Animal life is wonderfully diverse. Corals, starfish, **arachnids** (spiders) and humans are all animals. We consider ourselves special, but members of the animal kingdom make up less than 0.5% of the world's biomass.

The world's smallest animals are microscopic. The myxozoa, a group of tiny animals related to starfish, are only slightly bigger than a bacterium. Meanwhile the blue whale, which is the Earth's largest living animal, can grow up to approximately 30 metres (98 feet) long.

Most animals breathe oxygen, move around, reproduce sexually and develop from a small ball of cells called a blastula. Like fungi, animals are heterotrophs. This means that they cannot make their own food, and so have to acquire their nutrients externally. Unlike fungi, which absorb their food, animals obtain their nutrition by eating and digesting other organisms.

Animals are multicellular eukaryotes. They contain an enormous variety of specialised cells, for example muscle and neurons (nerve cells) (see page 30). These help them to perform complex functions, such as moving around and transmitting nerve impulses.

Animals vary dramatically in the way that they look, yet share a relatively small number of body layouts. Some animals, such as sea sponges, are asymmetrical, while others are symmetrical. Starfish, which belong to a group of animals called **echinoderms** are radially symmetrical, while beetles, which are **insects**, are bilaterally symmetrical. Most living animal species, including humans, show bilateral symmetry.

Animals can be categorised according to their ecological role. Carnivores eat meat, herbivores eat plants, omnivores eat everything, detritivores consume detritus and parasites feed on other organisms. Interactions between animals form complex food webs (see page 132).

Animal life evolved around 700 million years ago. The first animals were sea-sponge-like creatures. The first bilaterally symmetrical animals emerged

*Animals' bodies are diverse but they share a limited number of parts. This starfish is radially symmetrical, since it only has a top and bottom.*

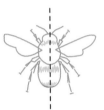

*Animals that have a front and a back as well as a top and a bottom are bilaterally symmetrical.*

70 million years later. These worm-like creatures were some of the world's first **invertebrates**.

During the Cambrian explosion, which started 540 million years ago, new varieties and body layouts emerged. These included snail-like **molluscs**, vertebrate animals and **arthropods**, which are animals like **crustaceans**, which have a segmented body and hard outer shell. Four hundred million years ago, as insects were evolving, the first four-legged animals crawled on land. In time, they gave rise to **amphibians**, **reptiles**, **birds** and **mammals**. Humans are a recent addition to the animal kingdom. *Homo sapiens* evolved approximately 300,000 years ago.

# Viruses

Viruses are very different to bacteria, archaea and eukaryotes. They are not made of cells. They are small packages of genetic information wrapped up in a protein coat called a capsid. Some are also surrounded by an outer envelope.

Viruses cannot replicate on their own. They can only replicate inside host cells. They do this by infecting the cell and hijacking its internal machinery. New copies of the virus are then released from the cell.

Viruses were first discovered in the 1800s by a Dutch scientist called Martinus Beijerinck. He demonstrated that a plant disease known as tobacco mosaic disease is caused by a tiny infectious agent. He named this pathogen 'virus', derived from the Latin word for venomous substance.

Since then scientists have described around 5,000 of the millions of viral species that are thought to exist. Viruses are found in almost every ecosystem on Earth, and are the most numerous biological entities.

These infectious agents cause disease in every type of living organism. Each virus can infect a range of hosts. Phages, for example, are viruses that infect bacteria. The tobacco mosaic virus affects a wide range of plants, while the poxviruses can infect vertebrates and insects.

Viruses spread in a variety of ways. Influenza viruses are transmitted by coughing and sneezing. HIV is mainly spread by sexual contact, while in the plant world sap-sucking insects often transfer viruses from one plant to another.

Scientists are unsure how and when they evolved, and how they are related to living species. This makes them difficult to classify. They are not viewed as part of the three-domain system.

### Influenza virus

*The influenza virus causes flu. Symptoms can be mild to severe, and sometimes pandemics break out. The Spanish influenza pandemic in 1918 killed 50 million people.*

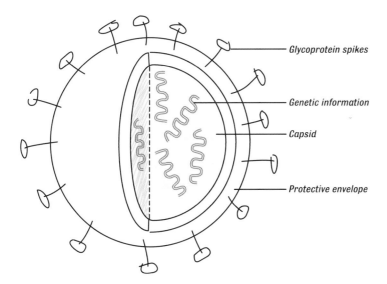

- Glycoprotein spikes
- Genetic information
- Capsid
- Protective envelope

# Reproduction

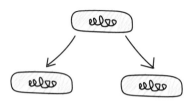

## Clones

*Bacterial cells divide and produce identical copies called clones.*

In order for life on Earth to continue, organisms must reproduce. There are two different forms of reproduction: sexual and asexual.

## Sexual reproduction

Sexual reproduction occurs in a wide range of living things, from single-celled organisms to humans. It involves two parents and specialised sex cells called gametes. In plants, the gametes are egg cells and pollen. In animals, the gametes are eggs (also called ova) and sperm. Gametes are formed by a special sort of cell division called meiosis (see page 35). The gametes from each parent meet and fuse. This is called fertilisation, and it leads to the creation of a zygote. The zygote then develops into a new individual.

Offspring created by sexual reproduction inherit genetic information from both parents. This makes them similar, but not identical, to their parents. This mixing up of genetic information leads to variation, which is an important component of evolution (see page 110). Variation helps species to adapt to change, so it is important for the long-term success of a species.

## Asexual reproduction

Asexual reproduction is common in the smallest animals and plants. It also occurs in bacteria, archaea, protists and fungi. It involves a single parent. There is no fusing together of specialised sex cells. Instead, the parent cell divides via a process called mitosis (see page 34).

The offspring that result are genetically identical to the parent. They are called clones. Because there is no mixing of genetic information, there is no variation amongst the offspring. This means that as the environment changes, they will be less likely to adapt and survive, compared with sexually reproducing species.

On the plus side, because asexual reproduction only requires one parent, it is efficient. Organisms don't need to spend time looking for a partner or having sex. Asexual organisms can produce large numbers of offspring quickly. Aphids, for example, can generate 18 generations in a single year without ever needing to mate.

Sometimes, larger plants and animals reproduce asexually. Tulips, potatoes and brambles can reproduce asexually, and, although rare, the phenomenon has also been documented in turkeys, hammerhead sharks, boa constrictors and Komodo dragons.

### Fertilisation in animals

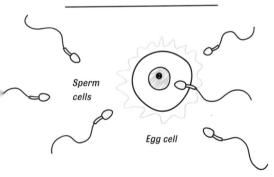

Sperm cells

Egg cell

Many organisms reap the benefits of sexual and asexual reproduction by doing both.

Most fungi reproduce asexually. The mushrooms that we eat are actually asexual fruiting bodies full of spores. The moulds that form on bread and other foods also reproduce asexually. When conditions are harsh – for example, if there is little water – then these fungi may switch to sexual reproduction. Hyphae from different fungi fuse to form a zygote, which is a genetic mixture of the two parents. The result is new individuals that are more varied, so they are better able to adapt to different environments.

Flowering plants, like strawberries and daffodils, reproduce sexually. The flowers contain the sexual organs, called stamens and carpels. The stamen is the male reproductive organ, which produces pollen. The carpel is the female reproductive organ. It contains the ovary, which houses the female gametes and the structure that produces them, called the ovule.

For sexual reproduction to occur, the pollen from one flower has to reach the ovules of another. This is called pollination. Sometimes the pollen is carried on the wind and sometimes it is transferred by pollinators, such as insects and birds. After pollination occurs, the resulting zygote develops into a seed, and the rest of the carpel becomes the fruit.

Many plants reproduce asexually. Strawberry plants produce specialised stems called runners that have tiny plants on them, while the tubers of daffodils and potatoes can be split to form new plants.

## Mammalian cloning

*Dolly the sheep was cloned from the DNA of a sheep mammary cell.*

Humans can clone organisms artificially. Some plants can be cloned by taking a cutting. Animal cloning is more difficult. Dolly the sheep was the first mammal to be cloned from an adult cell.

Researchers took DNA from an adult sheep cell and injected it into an egg cell that had had most of its DNA removed. The cell was coaxed to start dividing, and when the embryo was still tiny, it was transferred into the womb of a surrogate sheep, which then gave birth to Dolly.

Cloning is now used in farming. Scientists use cloning to make replicas of bulls that are valuable breeding animals.

## The reproductive organs of a flowering plant

*Most flowering plants grow bisexual flowers that have both male and female parts, but some plants have single-sex flowers that contain only male or female parts.*

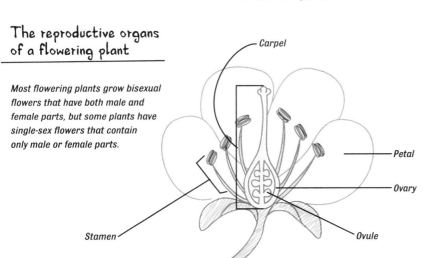

Carpel

Petal

Ovary

Stamen

Ovule

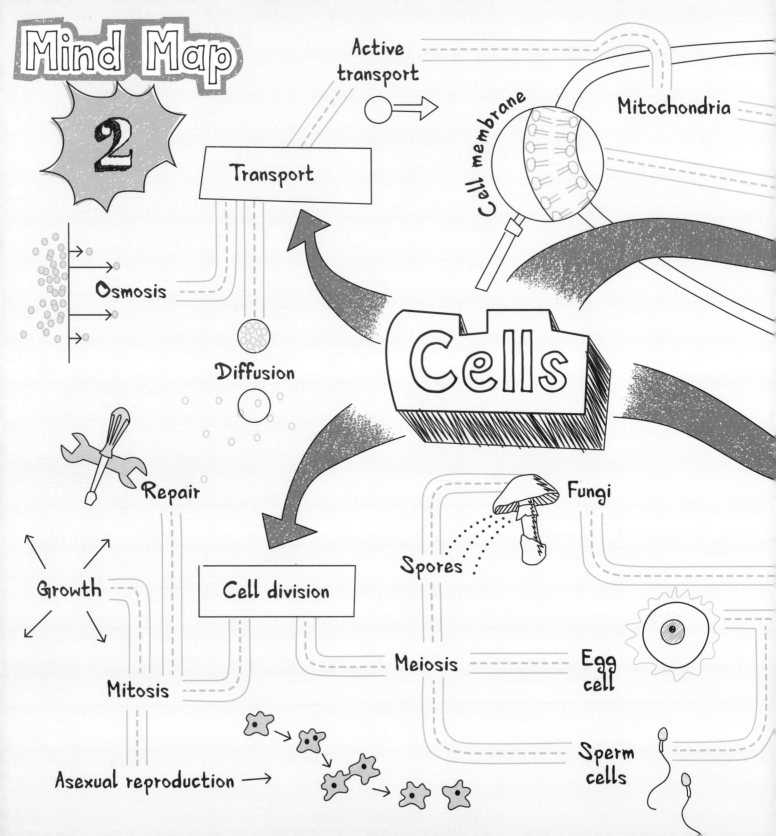

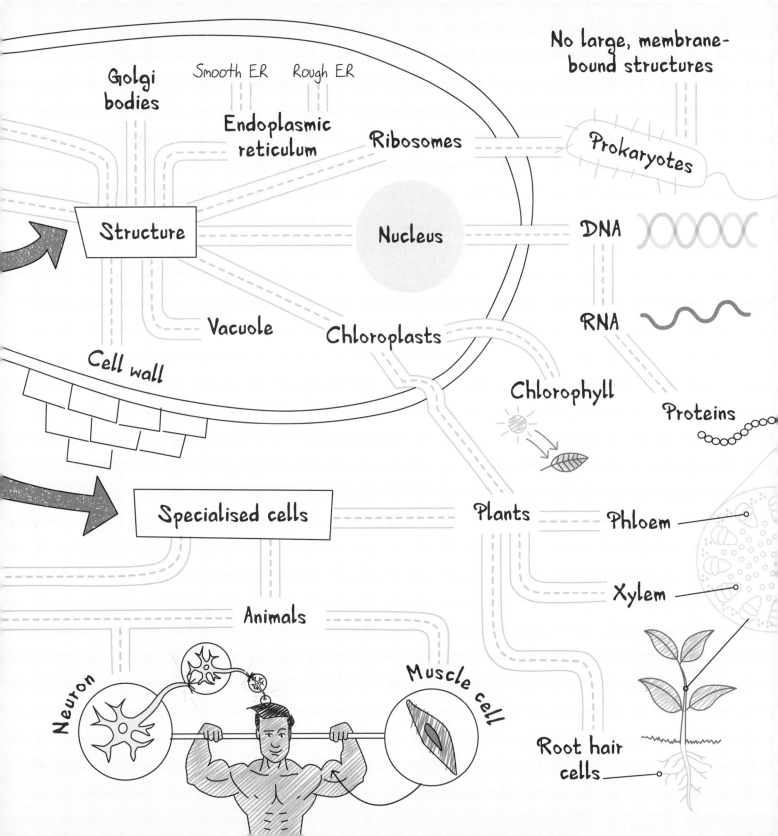

# The Molecules of Life

All living things are made of cells. They are the building blocks of life. Cells are made up of four basic molecules: carbohydrates, lipids, proteins and nucleic acids. Without these molecules, cells and organisms would not be able to live.

Macromolecules are large molecules that are made up of many smaller molecules. Carbohydrates, proteins and nucleic acids are all macromolecules.

## Carbohydrates

The simplest carbohydrates are monosaccharides, such as glucose and fructose. They form the basis of more complicated sugars such as polysaccharides. Polysaccharides, such as starch and glycogen, help cells store energy.

Plants store energy as starch. When the energy is needed, it can be broken down into glucose and used by the plant. Most animals have enzymes that break down starch, so they can unlock plant energy.

Animals store energy as glycogen. When it is broken down, energy is released, but it doesn't sustain us animals for long. If we don't eat, our glycogen stores become depleted in a day.

Polysaccharides are also used as building materials. Plant **cell walls** contain a polysaccharide called cellulose. Cellulose is the most abundant organic compound on Earth. Most animals are unable to digest cellulose, but cows contain microbes in their gut that do the job for them.

Chitin is another important carbohydrate. **Fungi** use it to build their cell walls, and arthropods use it to build their hard, outer shells.

## Lipids

Lipids, like fats, phospholipids and steroids, are important components of living cells.

Fat is another energy store. Mammals store their long-term food reserves in fat cells called adipose cells. Adipose tissue also helps to insulate the body, and forms a protective cushion around vital organs like the heart.

Phospholipids are important structural molecules. Phospholipids have water-attracting heads and water-repelling tails. They naturally assemble into a bilayer structure when there is water present. This bilayer structure forms the outer cell membrane of animal cells.

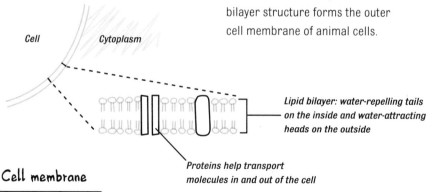

Cell      Cytoplasm

*Lipid bilayer: water-repelling tails on the inside and water-attracting heads on the outside*

Cell membrane

*Proteins help transport molecules in and out of the cell*

# Proteins

Proteins are large, complex molecules that are made up of smaller molecules called amino acids. Living things contain thousands of different proteins. They are enormously important, and have a wide range of functions.

Some proteins are enzymes. Enzymes are molecules that speed up chemical reactions. For example, digestive enzymes help animals to break down food (see page 44).

Some proteins are storage molecules. Casein is the main protein in mammalian milk. It stores amino acids so they can be used by baby mammals.

Hormonal proteins help to coordinate various activities. For example, when the hormone insulin is secreted by the pancreas it prompts other tissues to take up glucose. So, insulin regulates blood sugar levels.

Defensive proteins, like antibodies, help to detect and destroy viruses and bacteria, while transport proteins help to manoeuvre molecules around the body. For example, haemoglobin ferries oxygen from the lungs to other parts of the body.

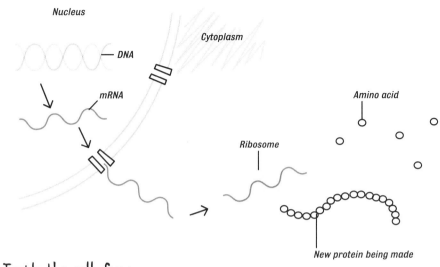

*Nucleus*

*DNA*

*Cytoplasm*

*mRNA*

*Amino acid*

*Ribosome*

*New protein being made*

## Inside the cell: from DNA to RNA to protein

*Genetic information controls cellular processes and guides the production of proteins. This involves multiple steps.*

Some proteins are involved in cell signalling. They recognise and respond to chemical signals, enabling cells to communicate with one another.

# Nucleic acids

Nucleic acids are long, thin molecules. **DNA** is a nucleic acid. It stores and transmits hereditary information. Organisms inherit their DNA from their parents (see page 96).

DNA is a set of instructions encoded in chemical form. These instructions guide everything that goes on inside a cell. Sequences of DNA called genes are code for the production of proteins. This is a multi-step process and it involves a second nucleic acid called **RNA**.

RNA is very similar to DNA. A special sort of RNA called messenger RNA (mRNA) helps the cell to read the instructions that are written in nuclear DNA, and then convert them into proteins in the cytoplasm (see page 99).

**Prokaryotes** and eukaryotes all store their genetic information as DNA, but some viruses store theirs as RNA.

# Eukaryotic Cells

## Animal cells

Animal cells come in many shapes and sizes. A human **egg cell** is a spherical structure about 0.1 millimetre (mm) (0.004 inches) in diameter. It is just big enough to be seen with the human eye.

Most animal cells are much smaller than this. They are miniscule. The only way to see them is with a microscope.

The first light microscope was developed over 300 years ago. Light microscopes use a beam of light to form a magnified image of an object. The most powerful light microscopes can magnify images up to 2,000 times.

Since then, new microscopes have been invented. Electron microscopes use a beam of electrons to form an image and can magnify objects up to 2 million times. Confocal microscopes enable scientists to reconstruct detailed 3-D images of the inside of cells. These new microscopes are much more complicated and expensive than light microscopes, which can be found in schools and laboratories all over the world. Microscopy has revealed the **structure** of many cells.

The **nucleus** is the cell's control centre. It is a small membrane-bound structure. It contains the vast majority of the cell's DNA. An average human cell contains 2 metres (6.5 feet) of DNA all coiled up.

The cytoplasm forms the bulk of the cell. It is the gel-like substance that contains the nucleus and all the other structures that are found inside a cell. It is the place where most cellular activities occur, including cell division and important chemical reactions.

The **cell membrane** is the outer layer that surrounds the cytoplasm. It is made of two layers of phospholipids, interspersed with much larger proteins. The cell membrane controls the movement of substances in and out of the cell.

**Mitochondria** are tiny, battery-like structures found inside the cytoplasm. They provide the cell with energy, which they make by respiration. They are minute and contain a tiny amount of DNA.

**Ribosomes** are tiny, protein-making factories. They link together amino acids to form proteins. Cells that make lots of **proteins** have many ribosomes. Human pancreas cells make lots of digestive enzymes, so each cell contains millions of ribosomes.

The **endoplasmic reticulum** (ER) is a network of flattened, tube-like structures. There are two different types. **Rough ER** is studded with ribosomes, and so is involved in making proteins. **Smooth ER** lacks ribosomes and is involved in making lipids.

**Golgi bodies** are like warehouses that receive, make, modify and ship all sorts of different cellular products, such as enzymes and other proteins. Golgi bodies are made up of a collection of flattened membranous sacs, so they look like a stack of mini pitta breads. Owing to their large, unusual shape, they were one of the first cellular structures to be identified. Italian physician Camilo Golgi discovered them in 1898 when he was studying the nervous system.

## Animal cell

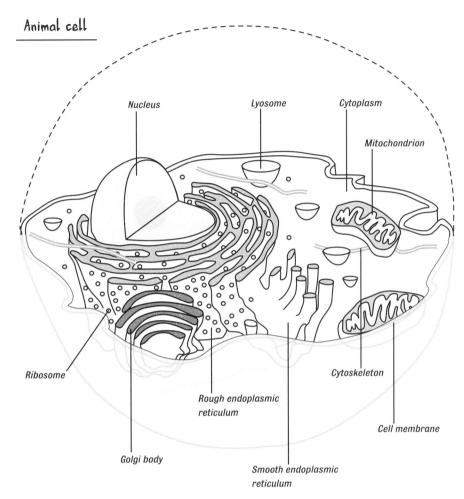

Nucleus

Lyosome

Cytoplasm

Mitochondrion

Ribosome

Golgi body

Rough endoplasmic
reticulum

Smooth endoplasmic
reticulum

Cytoskeleton

Cell membrane

Lysosomes are tiny, membrane-bound sacs. They contain enzymes that break down big molecules into smaller ones. They help to digest nutrients; producing sugars, amino acids and other molecules that can be used by the cell. They also help to break down and recycle damaged organelles that are no longer needed.

The membrane-bound structures of eukaryotic cells are called organelles. For a long time, scientists thought they floated freely in the cytoplasm, but then microscopy showed that they are embedded in a network of supportive fibres. This is called the cytoskeleton.

## SPECIALISATION IN ANIMAL CELLS

Adult animals contain hundreds of kinds of **specialised cells**, all adapted to perform specific functions. Some, like eggs and sperm cells, work on their own. Others, like neurons (nerve cells) and **muscle cells**, are adapted to work as part of a tissue, an organ, or a whole organism.

**Sperm cells** contain genetic information from the male parent. They are mobile cells that must travel in order to find and fertilise an egg. Sperm cells have several adaptations to help them achieve this.

Each sperm cell is small and thin with a long tail that whips from side to side, helping to propel it along. They are packed with mitochondria, to give them the energy that they need for swimming.

## Sperm cell

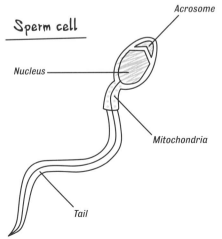

Acrosome

Nucleus

Mitochondria

Tail

The head of the cell contains the DNA-containing nucleus and a specialised structure called an acrosome. The acrosome contains digestive enzymes that help to break down the outer layers of the egg so the sperm can penetrate inside (see page 29).

**Neurons** are specialised to transmit and receive information. They are found in the brain and spinal cord, and in the cable-like bundles of nerve fibres that are spread throughout our bodies. The human brain contains an estimated 100 billion neurons.

Neurons receive signals from other cells via branches called dendrites. A single neuron may have hundreds of dendrites.

They also have long projections called axons. The longest axon in your body runs from the base of your spine to the tip of your big toe. Most neurons only have one axon.

Some axons are surrounded by an outer layer called the myelin sheath. This helps electrical information to travel swiftly along the axon.

Where the dendrite of one cell meets the axon of another, there is a tiny gap called a synapse (see page 71).

## Plant cells

Plants and animals are very different, yet their cells share many common features. Like animal cells, plant cells are made up of an organelle-packed cytoplasm. This includes a DNA-filled nucleus, endoplasmic reticulum, Golgi bodies and energy-generating mitochondria.

Plants and animals have very different lifestyles. Most animals are mobile, while plants are unable to get up and move around. And while animals consume nutrients from elsewhere, plants make their own via photosynthesis. As a result, plant cells also have a number of unique features.

Plant cells contain **chloroplasts**. They are tiny structures that contain a green substance called chlorophyll. Chloroplasts are found in all the green parts of a plant, such as the leaves, shoots and stems. They are where photosynthesis occurs. However, because roots are underground they do not contain chloroplasts, and so do not photosynthesise.

**Chlorophyll** is a pigment that absorbs light so the plant can make food via photosynthesis.

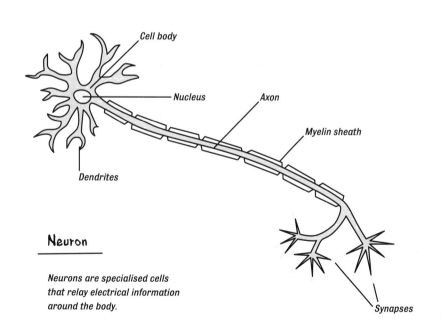

Cell body

Nucleus

Axon

Myelin sheath

Dendrites

## Neuron

*Neurons are specialised cells that relay electrical information around the body.*

Synapses

## Plant cell

Vacuole

Mitochondrion

Nucleus

Ribosome

Cytoplasm

Chloroplast

Golgi body

Cell wall

Smooth endoplasmic reticulum

Rough endoplasmic reticulum

## SPECIALISATION IN PLANT CELLS

Plants and animals live very different lives, so their cells are specialised in different ways. Photosynthetic cells are specialised to perform photosynthesis, but there are many other types of specialised plant cells, such as root hair cells, xylem cells and phloem cells.

**Root hair cells** are found at the end of growing roots. They are specialised to help plants take up water and mineral ions. They have long, hair-like structures that stick out from the cell body. This increases the surface area that is available for water to move into the cell.

They have thin cell walls and no thick, waxy outer layer. This helps the cell to take up water more quickly. They are packed with mitochondria, which release the energy that is needed for cells to take up the mineral ions they need.

Plant cells also contain a permanent **vacuole**. The vacuole is a large space in the middle of the cytoplasm that is filled with sap. It helps to keep the cells rigid and support the plants.

Instead of being surrounded by a flexible cell membrane, plant cells are surrounded by a rigid cell wall. This is much thicker than the cell membrane that surrounds an animal cell. The cell wall helps to protect the plant, maintain its shape and prevent it from taking in too much water.

Although the makeup of this cell wall varies from species to species, all plant cell walls are strengthened with the polysaccharide, cellulose.

## Root hair cell

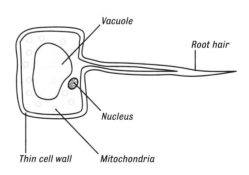

Vacuole

Root hair

Nucleus

Thin cell wall

Mitochondria

## Xylem cells

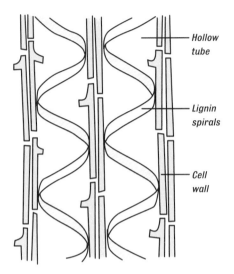

- Hollow tube
- Lignin spirals
- Cell wall

## Phloem cells

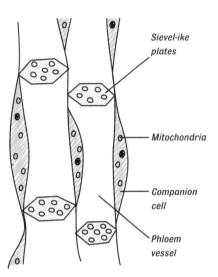

- Sievel-ike plates
- Mitochondria
- Companion cell
- Phloem vessel

## Algae

Algae are eukaryotic life forms (see page 16). They live in water, and vary massively in shape and size. Algae range from single-celled life forms like diatoms to big, multicellular organisms like giant kelp, which can grow up to 30 metres (98 feet) long.

For a long time, scientists classified algae as plants, but now they realise algae belong to the kingdom of protists (see page 17). Just like plants, algae make their own food by photosynthesis, so their cells contain chloroplasts. Their cell walls are also strengthened by cellulose.

**Xylem** is a type of transport tissue found in plants. Xylem tissue carries water and mineral ions from the roots to the leaves.

When they are first formed, xylem cells are alive. Over time, a chemical called lignin forms hollow spirals inside the cells. The cells die, leaving long, hollow tubes that water and mineral ions can pass through.

We call lignified cells 'wood'. The spirals of lignin are very strong, so they also help the plant to stand up.

**Phloem** is also a type of plant transport tissue. It distributes the food that the plant makes. It is made of phloem cells that form tubes, but unlike xylem cells, the phloem cells do not become lignified and die.

To help these nutrients move around more freely, the walls between phloem cells break down to form sieve-like plates. To make more room for the nutrients to move, phloem cells lack many basic structures, including nuclei and ribosomes. To help them survive, they are supported by neighbouring companion cells.

## Algae cells

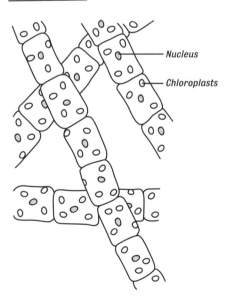

- Nucleus
- Chloroplasts

# Prokaryotic Cells

Prokaryotic cells, such as bacteria and archaea, differ from eukaryotic cells because they have **no large, membrane-bound structures** in their cytoplasm. There are no nuclei, mitochondria or chloroplasts. Bacterial cells have tangled circles of DNA that float freely in the cytoplasm.

Like all other organisms, bacteria contain ribosomes that make proteins, but the structure of the bacterial ribosome is different to that of eukaryotes and archaea.

The cytoplasm is surrounded by a cell membrane made of phospholipids, and the cell membrane is surrounded by an exterior cell wall. This is different from plant cell walls that contain cellulose, and fungus cell walls that contain chitin. Bacterial cell walls are made of a complex molecule called peptidoglycan.

The cell wall is vital for the survival of many bacteria. The antibiotic penicillin is able to kill some bacteria because it prevents peptidoglycan from being made.

Bacteria can be classified into gram-positive and gram-negative varieties, based on the structure of their cell walls. Gram-positive bacteria have a thick cell wall containing many layers of peptidoglycan. Gram-negative bacteria have a relatively thin cell wall with fewer layers of peptidoglycan.

Rod-shaped bacteria, such as *Salmonella*, also have a protective slime layer that covers the cell wall, and long, mobile tails called flagella, which they use to move around.

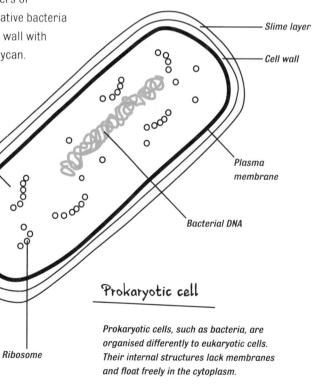

Slime layer

Cell wall

Plasma membrane

Bacterial DNA

Cytoplasm

Flagellum

Ribosome

### Prokaryotic cell

*Prokaryotic cells, such as bacteria, are organised differently to eukaryotic cells. Their internal structures lack membranes and float freely in the cytoplasm.*

# Cell Division

In eukaryotic cells, the DNA inside the nucleus is split into smaller, more manageable chunks called chromosomes. Human cells contain 23 pairs of chromosomes (see page 102).

Chromosomes are long segments of DNA, peppered with genes. Genes contain the information needed to make new cells, tissues and organisms.

## Mitosis

Sometimes the body needs to produce identical copies of cells that already exist. This is done by a process called mitosis. The cells that are produced have the same chromosomes, and the same genetic information, as the parent cells.

### Cells divide

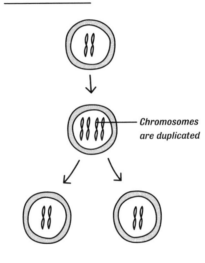

Chromosomes are duplicated

*Daughter cells are identical to the original.*

In multicellular organisms, mitosis is used to create the cells that are needed for **growth** and **repair**. Unicellular organisms use mitosis for **asexual reproduction** to make identical copies of themselves. Actively dividing cells pass through a series of stages known collectively as the cell cycle. Mitosis is part of the cell cycle.

## THE CELL CYCLE

The cell cycle can be long or short. The length of the cell cycle depends on the cells that are involved and the life stage of the organism. For example, the cells in an adult tend to divide more slowly than the cells of an embryo. There are exceptions, including skin cells, blood cells and hair follicle cells that divide regularly through life.

The cell cycle follows a number of stages that always occur in order. In the first stage, the cell grows bigger and carries out its normal activities, but it also starts getting ready to divide. The cell makes duplicate copies of all of its chromosomes, as well as extra organelles, such as chloroplasts and mitochondria. This is the longest phase.

During the second stage, the cell begins to divide. This is mitosis. The two sets of chromosomes are pulled to opposite sides of the dividing cell, and the nucleus splits in two. In the final stage, the cytoplasm and cell membranes divide to produce two identical daughter cells.

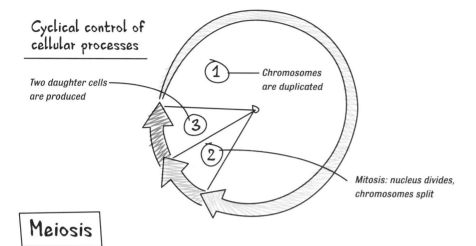

## Cyclical control of cellular processes

1. Chromosomes are duplicated

Two daughter cells are produced

3

2

Mitosis: nucleus divides, chromosomes split

## Meiosis

Meiosis is a special form of cell division that is used to make sex cells, such as sperm, eggs and **spores**. Unlike mitosis, which produces daughter cells with the same number of chromosomes as the parents, meiosis produces cells with half the original number of chromosomes.

Most cells in the body contain two sets of chromosomes; one from each parent. Sex cells are specialised cells that contain one set of chromosomes. Then, when the sex cells meet and fertilisation occurs, the resulting embryo has two sets of chromosomes; one from each parent.

In animals, the female sex cells (egg cells) are made in the ovaries, and the male sex cells (sperm) are made in the testes. Meiosis occurs in both these places.

Meiosis is similar to mitosis, but it contains an additional round of cell division. The original cell has two sets of chromosomes. These chromosomes are copied, so the cell now contains four sets. It then divides twice, in quick succession, to form four daughter cells. These daughter cells contain a single set of chromosomes.

All of the daughter cells are genetically different to one another. This is because the chromosomes become mixed up during meiosis. Sections of DNA get swapped. This is good because it introduces variation. More variation occurs after fertilisation, when the chromosomes from the separate sex cells combine to form a complete, double set.

Different organisms have different numbers of chromosomes. Human eggs have 23 single chromosomes, as do human sperm. When the egg is fertilised, the resulting cell has 23 pairs of chromosomes. From now on, the cell and its progeny divide by mitosis to form a new individual. The number of cells increases rapidly. As the embryo develops, some of the cells divide to produce more specialised forms, which go on to produce tissues, organs and organ systems (see page 41).

## Making sex cells

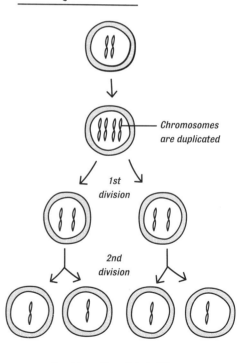

Chromosomes are duplicated

1st division

2nd division

*Daughter cells contain half the original number of chromosomes.*

# Transport

Cells need to be able to take in some substances, like nutrients and oxygen, and get rid of others, like waste products. Dissolved substances can move in and out of the cell via the cell membrane. There are different ways of doing this.

## Diffusion

Diffusion is the movement of particles from an area that is more concentrated to an area that is less concentrated. It happens because particles move around randomly and bump into each other. This makes them spread out.

Simple sugars such as glucose, gases such as oxygen and carbon dioxide, and waste products such as urea, all move around by diffusion. When you breathe, the oxygen you inhale travels to your lungs. It then moves into your red blood cells by diffusion. It moves from an area of high oxygen concentration in your lungs, to an area of lower oxygen concentration in your blood.

When cells respire (see page 60), they produce carbon dioxide as a waste product. This builds up inside the cells, until the concentration of carbon dioxide inside the cell exceeds the concentrations of carbon dioxide outside the cell. It then diffuses out through the cell membrane and into the blood.

The liver makes urea when it breaks down proteins. This waste product then moves out of the liver cells by diffusion.

## Osmosis

Osmosis is a special sort of diffusion. Solutions that are inside cells are separated from the outside by the cell membrane. This membrane only lets certain molecules through, so it is called a selectively permeable membrane. Osmosis is the diffusion of water molecules from an area that has a high concentration of water to an area with a lower concentration of water, across a selectively permeable membrane. It is very important because it helps cells to regulate the concentrations of vital molecules.

### Diffusion of particles

*The black particles spread through the liquid by diffusion. They move from an area of high concentration to areas of lower concentration.*

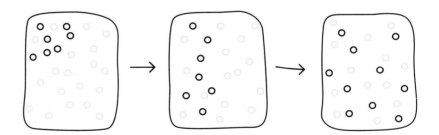

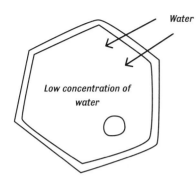

*Water*

**Low concentration of water**

**High concentration of water**

## Osmosis

*Water enters plant cell by osmosis. This helps keep the cell turgid and firm.*

Plants rely on osmosis to keep them standing upright. Water moves into plant cells by osmosis. This causes the vacuole to swell and puts pressure on the plant cell wall. This makes the cells firm, which makes the leaves and stems firm too.

Similarly, osmosis helps animal cells to regulate the concentration of their contents. If a cell uses up water and its cytoplasm becomes too dense, then water moves into the cell by osmosis. If a cell contains too much water, then water leaves by osmosis. In both cases, osmosis restores the natural balance.

# Active transport

Sometimes, cells need to transport dissolved substances from an area of low concentration to an area of high concentration. They do this by active transport. This is the opposite of diffusion and osmosis. Unlike diffusion and osmosis, which do not require energy, active transport does require a source of energy. Cells produce this energy during respiration.

## Active transport in a shark

*Salt in the shark's blood is manoeuvred into the salt gland by active transport, then expelled from the body back into the water.*

Active transport is also unique because it requires a carrier protein. Carrier proteins are molecules that span the cell membrane. During active transport, these carrier proteins pick up specific molecules and transport them through the membrane.

Active transport is important because it helps organisms to move molecules around internally. In our bodies, active transport is used to move sugars like glucose out of our gut and kidneys, and into our blood. It lets cells absorb ions from very dilute solutions. For example, plants use active transport to absorb minerals, like nitrates, from the soil. Sharks that live in salty water use active transport to help them remove excess salt from their bodies. They have special salt glands in their rectum, which help to move salt from the blood into the gland, where it is then excreted as a concentrated solution.

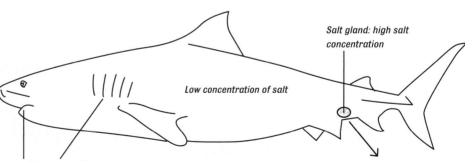

*Salt gland: high salt concentration*

*Low concentration of salt*

*Salt enters the shark's body via the mouth and the gills*

*Salt is then excreted from this special gland*

# Mind Map 3

## Organising Life

Circulatory

Red blood cells

White blood cells

Cell types

Systems

Digestive

Mouth

Oesophagus

Gall-bladder

Stomach

Respiratory

Lungs

Double circulatory system

Heart

Organs

Anus

Rectum

Small intestine

Large intestine

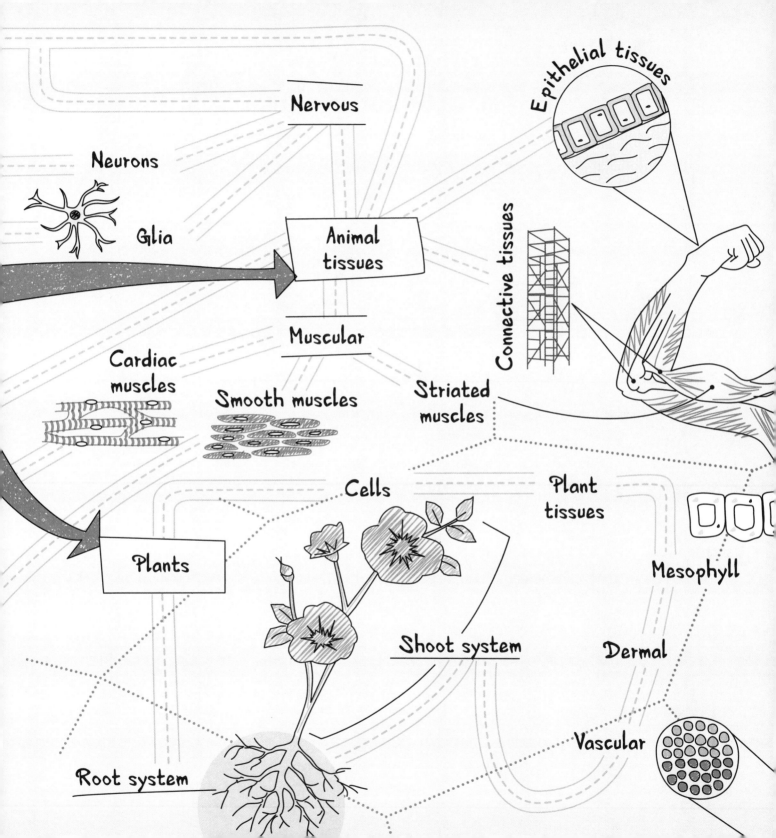

Nervous

Neurons

Glia

Animal tissues

Epithelial tissues

Connective tissues

Muscular

Cardiac muscles

Smooth muscles

Striated muscles

Cells

Plant tissues

Plants

Mesophyll

Shoot system

Dermal

Root system

Vascular

# Animal Growth and Development

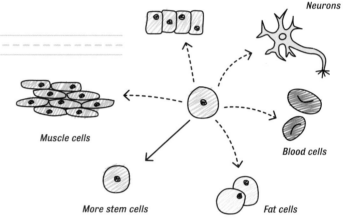

Epithelial cells

Neurons

Muscle cells

More stem cells

Blood cells

Fat cells

## Stem cells

*Stem cells are multi-talented. They can make copies of themselves as well as other, more specialised cells.*

When life begins, and the embryo is nothing but a small ball of rapidly dividing cells, all of the cells are similar. These cells are called stem cells. Stem cells are unspecialised cells. They can divide and make extra copies of themselves, but they can also produce more specialised cells, such as muscle cells and neurons. When a stem cell produces a more specialised cell, it is called differentiation.

In animals, specialised cells arise very quickly. By the time a human baby is born, most of its cells are already highly specialised.

Mitosis fuels the growth of living things, but when an animal matures, it stops growing. After that, when mitosis does occur it tends to be to replace damaged or dying cells (see page 34).

Although the cells inside an organism all share the same genetic code, different genes are switched on and off in different cell types. We say these cells have different patterns of gene expression. As a result, the pattern of gene expression in a neuron is different to the pattern of gene expression in a muscle cell. This is what gives the neuron its specialised form and function.

Most specialised cells can only divide by mitosis. This means they are unable to generate different types of specialised cell. For example, when a muscle cell divides, it can only produce muscle cells.

Some specialised cells cannot divide. Red blood cells, for example, wear out all the time. New red blood cells are produced by stem cells that live in the bone marrow. The human body makes about 2 million new red blood cells every second.

**Neurons** are also unable to divide. Although there are some stem cells in the adult brain, most of the neurons that die are not replaced. Neuroscientists are trying to find ways to activate the brain's stem cells, to help slow the progress of degenerative conditions such as Alzheimer's disease.

# From Cells to Systems

Single-celled organisms can easily exchange molecules with the environment. An amoeba in a pond can absorb nutrients directly from the water. Oxygen and nutrients diffuse in across the cell membrane, and carbon dioxide and waste products diffuse out.

## Amoeba

*Single-celled organisms such as amoebae exchange substances directly with the environment.*

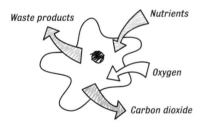

Waste products · Nutrients · Oxygen · Carbon dioxide

It's more difficult for multicellular organisms to exchange molecules with the environment. The adult human body contains an estimated 37,000,000,000,000 cells. Most of these have no contact with the outside world, so they can't exchange substances like nutrients and waste directly.

Instead, the cells of complex organisms are surrounded by an internal environment of extracellular fluid. Humans are 60% water. Two-thirds of this is the fluid found inside cells, and one-third is the 'extracellular' fluid that surrounds cells. So, instead of exchanging molecules with the external environment, the cells of complex organisms exchange molecules with this extracellular fluid.

All of the cells in an organism work together in a coordinated manner to perform all of the processes that keep us alive. Cells are organised into tissues, organs and organ systems.

Tissues are groups of cells that look alike and share similar functions. For example, muscular tissue helps us to move and nervous tissue helps to spread information.

**Organs** are structures made up of two or more tissues that are organised to serve a particular function. So, the heart pumps blood, the lungs bring oxygen into the body, and the stomach helps to digest food.

Organ systems are groups of organs that work together to perform specific functions. So, the nervous system, which includes the brain, spinal cord and various nerves, helps to respond to stimuli and coordinate bodily activities.

The various systems depend on one another. For example, the cells of the nervous and digestive systems depend on oxygen from the respiratory system, and the cells of the respiratory system require nutrients from the digestive system.

## From cells to systems

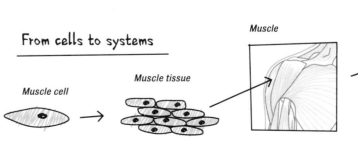

Muscle cell · Muscle tissue · Muscle · Muscle system

# Animal tissues

Large multicellular organisms, such as humans, are made up of four basic types of tissue: nervous tissue, connective tissue, muscle tissue and epithelial tissue.

**Epithelial tissues** are sheets of cells that cover the outside of the body, and line organs and body cavities inside the body. The epidermis, which is the outermost layer of the skin, is an epithelial tissue. So are the cells that line the nasal passage.

The cells of the epithelium are tightly packed together, so they provide a barrier against injury, infection and fluid loss.

Epithelial cells are polarised because they have a top and bottom side. The top or 'apical' side faces the cavity or outside of the organ, so it is exposed to fluid or air. Often the apical side is covered in specialised projections. The apical side of the epithelial cells that line the small intestine are covered in tiny, finger-like structures that increase the surface area for absorbing nutrients. The opposite side of the epithelium is the basal surface.

## Epithelial tissue of the small intestine

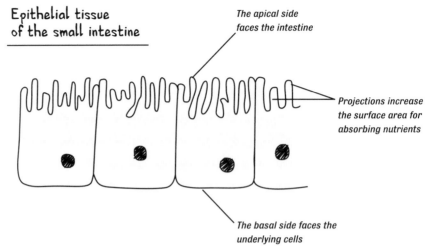

*The apical side faces the intestine*

*Projections increase the surface area for absorbing nutrients*

*The basal side faces the underlying cells*

**Connective tissues** are supportive tissues. Unlike epithelial tissue, which has cells that are closely packed together, connective tissue contains sparsely scattered cells. The cells are suspended in a gel-like substance that contains crisscrossing fibres made of protein.

Loose connective tissue is the most common type of connective tissue. It is found throughout the body, where it helps to hold organs in place and binds epithelial tissues to the underlying tissues.

## Loose connective tissue

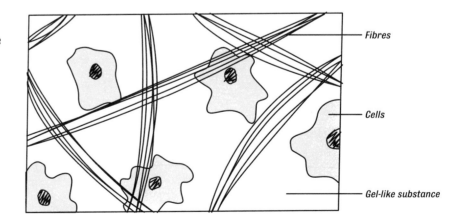

*Fibres*

*Cells*

*Gel-like substance*

Dense or fibrous connective tissue is packed full of collagen fibres. It is found in the tendons that connect muscles to bones, and the ligaments that connect bones to each other.

Adipose tissue, bone, cartilage and blood are all types of specialised connective tissue. Cartilage, for example, is a strong, flexible material that protects the end of certain joints. It also helps to keep the airways open and cushions the discs that separate the bones of the spinal column.

**Nervous** tissue is involved in receiving, processing and transmitting information. It is made up of neurons, which transmit information, and other cells, known as **glia**.

For years, scientists thought of glia as structural cells that physically support the nervous system. It's now clear that they have many roles, including destroying pathogens and supplying the neurons with oxygen and nutrients.

**Muscular** tissue allows the body to move. Muscle cells contain protein filaments that enable the muscles to contract. Vertebrates have three different types of muscle.

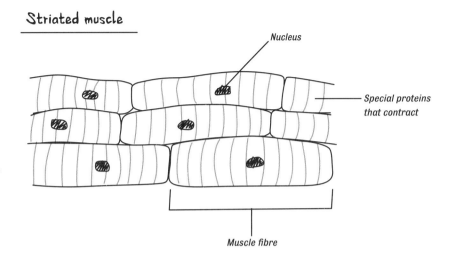

Striated muscle

*Nucleus*

*Special proteins that contract*

*Muscle fibre*

**Striated muscles**, also called skeletal muscles, are attached to bones by tendons. They work in pairs to move the bones: when one muscle in a pair relaxes, the other contracts. Skeletal muscles allow you to consciously control your movements.

Striated muscles are made of bundles of long cells called muscle fibres. It takes energy to make a muscle work, so muscle fibres contain lots of mitochondria. Striated muscle is so called because it looks stripy.

**Smooth muscles** are not stripy. Smooth muscles control the movements that are not under conscious control. You don't have to concentrate to squeeze food along your gut; instead, your smooth muscles do it for you. Smooth muscle is found in the walls of the bladder, blood vessels and other internal structures.

**Cardiac muscles** are only found in the walls of the heart. They are stripy like striated muscle, but they are not under voluntary control, so you don't have to remember to make your heart beat. The individual fibres are connected by structures called intercalated discs, which relay signals from cell to cell, allowing them all to contract in sync.

# The Human Digestive System

The food we eat has to be broken down into smaller molecules that our bodies can use. This is called digestion. Digestion is important because without it we would be unable to absorb nutrients from food.

The adult human **digestive** system is about 9 metres (29.5 feet) long. It starts at one end with the **mouth**, and finishes at the other with the **anus**. Somewhere in between, a complex variety of organs, tissues and cells work together to help you digest your food. It is one of the most important organ systems in your body.

Digestion begins before you even take a bite. The sight and smell of food is enough to prompt the stomach to start making digestive enzymes. This prepares the digestive system for incoming food.

In the mouth, the food is chewed and mixed with saliva. This is the mechanical part of digestion. It breaks the food into smaller pieces, which can be swallowed and digested more readily.

## Human digestion

*The human digestive system contains many different organ, tissue and cell types.*

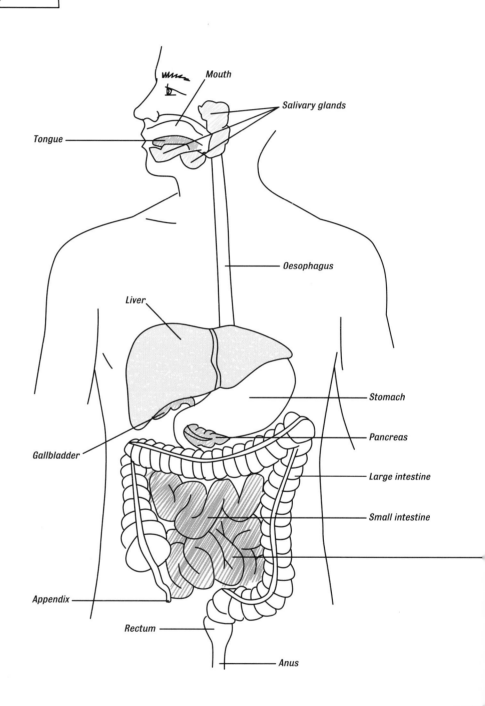

Mouth

Salivary glands

Tongue

Oesophagus

Liver

Stomach

Pancreas

Gallbladder

Large intestine

Small intestine

Appendix

Rectum

Anus

The food then enters the **oesophagus**, where waves of smooth muscle contraction push it along and into the stomach.

The **stomach** is located just above the diaphragm. It has two important jobs. The first is to store food. The stomach is stretchy and can hold about 2 litres (68 fluid ounces) of food and fluid. The second is to chemically digest the food.

Cells in the stomach secrete a digestive fluid called gastric juice, which mixes with the food as it is churned about. Gastric juice contains acid and enzymes (see page 56), which help to break large insoluble food molecules down into smaller, more soluble ones.

Digestion then continues in the **small intestine**. This is the longest part of the digestive system. It is a convoluted tube about 6 metres (20 feet) long. In the first part of the small intestine, the food mixes with digestive juices from the pancreas, liver and **gallbladder**, as well as from glands in the wall of the intestine itself. The liver, for example, releases bile, which helps to break down otherwise indigestible fat molecules.

With digestion now largely complete, food molecules continue along the small intestine, which is lined with big, wiggly folds called villi. The villi are covered in tiny finger-like projections called microvilli. Together, these folds have a surface area of around 250 square metres (2,700 square feet), which is about the same size as a tennis court.

This expanded surface increases the rate at which nutrients can be absorbed into the bloodstream. Some nutrients enter the bloodstream by diffusion, while others are manoeuvred by active transport.

The small intestine joins the **large intestine** at a T-shaped junction. One arm of the 'T' is the colon, which is about 1.5 metres (5 feet) long, which leads to the **rectum** and anus, and the other is a pouch called the cecum. In humans, the cecum includes a tiny dead end called the appendix.

For many years, scientists thought the appendix was little more than an evolutionary relic, but now it's thought the structure contains a reservoir of beneficial bacteria that may help to fend off harmful microbes.

In herbivores like cows, the cecum is much larger. It is a place where undigested plant matter is fermented, so the cellulose it contains can be broken down into a more usable form.

As food enters the large intestine, most of the available nutrients have already been removed from it. In the colon, any remaining nutrients and excess water are absorbed, before the indigestible leftovers are sent to the rectum.

These waste products are then stored in the rectum, awaiting a convenient moment to expel them via the anus.

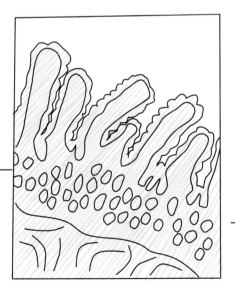

### Small intestine

*Tiny folded projections vastly increase the surface area of the small intestine.*

# The Circulatory and Respiratory Systems

The **circulatory** system is formed of the heart, blood and blood vessels. It transports hormones, removes waste products, and delivers oxygen and nutrients around the body.

Blood is a unique tissue. It is made from a variety of **cell types**, including red and white blood cells. These float freely in a yellowish liquid called plasma. The plasma is part of the extracellular fluid. **Red blood cells** pick up oxygen from the lungs and transport it to the cells where it is needed. Red blood cells contain haemoglobin, an iron-containing molecule that binds oxygen. Haemoglobin gives red blood cells their red colour.

**White blood cells** are part of the body's immune system. Some make antibodies, while others help to destroy invading bacteria and viruses. Blood is transported around inside tubes called blood vessels. There are three main types of blood vessel.

**Artery**
*(thick walls)*

**Vein**
*(thinner walls)*

**Capillary**
*(thinnest walls)*

### Blood vessels

*Blood travels around the body in a variety of different vessels.*

Arteries carry red, oxygenated blood away from the heart toward the organs. Arteries have thick, muscular walls. Veins carry deoxygenated blood away from the organs toward the heart. Veins are thinner than arteries, and have valves to prevent blood from flowing backward in the wrong direction.

Capillaries are a huge network of tiny blood vessels. Their walls are even thinner. This allows useful substances like oxygen and glucose to diffuse out of the blood and into the body's cells, and waste products, like carbon dioxide, to pass from the body into the blood.

Blood vessels form a network that spans the entire body. They connect different tissues, organs and organ systems together. In mammals, the blood vessels are arranged into a **double circulatory system**. One system carries blood from the heart to the lungs and back again, and the other carries blood from the heart to all of the other organs in the body and back again. This is an efficient system because it enables oxygen-rich blood returning from the lungs to be pumped around the body quickly.

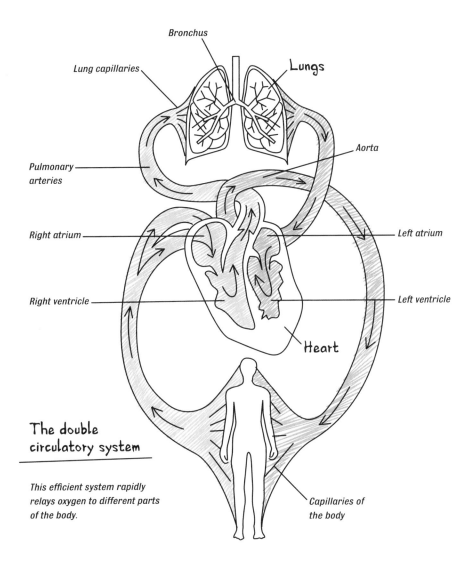

Bronchus

Lung capillaries

Lungs

Pulmonary arteries

Aorta

Right atrium

Left atrium

Right ventricle

Left ventricle

Heart

The double circulatory system

*This efficient system rapidly relays oxygen to different parts of the body.*

Capillaries of the body

The ventricles contract to force blood out of the heart. The right ventricle pumps deoxygenated blood to the lungs. The left ventricle pumps oxygenated blood around the body in a big artery called the aorta. The wall of the left ventricle is much thicker than the wall of the right ventricle. This is because it has to pump blood around the whole body, rather than just to the lungs.

The **lungs** are the major organ of the **respiratory** system. The average person takes around 25,000 breaths every day. Every time you breathe in, oxygen passes down the windpipe into two bronchi, which lead into the lungs. Each bronchus divides into smaller bronchi, which divide repeatedly into smaller tubes called bronchioles. At the end of the bronchioles are air sacs called alveoli. They provide a large surface area for gas exchange to occur.

The oxygen we breathe in diffuses across the alveoli into the blood, and the carbon dioxide that builds up in the body diffuses out of the blood and into the alveoli. The carbon dioxide is expelled when we exhale.

All of the body's organ systems are intricately interconnected, and although they are built from a limited number of cell and tissue types, they work together efficiently to perform the tasks needed to keep you alive.

The **heart** is a vital organ because it pumps blood around the body. It is made of two pumps that drive the double circulation. The pumps beat together at a rate of around 70 beats per minute. The heart is made up of four chambers. Blood enters the top chambers, called atria. Blood entering the right atrium is deoxygenated blood that comes from the body. Blood entering the left atrium is oxygenated blood that comes from the lungs.

The atria contract together forcing blood down into the bottom chambers, which are called ventricles. Valves close to stop the blood from flowing back up into the atria.

# Plant Growth and Development

A big difference between plants and animals is that most plants keep growing throughout their lives. Plants can keep growing because they have growth zones called meristems. The meristems can be found at the tips of roots and shoots.

The meristems contain stem cells that divide by mitosis to produce lots of new cells, which then expand and help the plant to grow. Some of the new cells remain in the meristem and keep producing new cells. Others differentiate into more specialised cell types, such as epidermal cells and root hair cells (see page 31). This is called primary growth.

Primary growth enables roots and shoots to grow longer. In non-woody plants, such as herbs and ferns, it produces almost all of the plant body.

Woody plants, like trees, also grow in circumference. This is called secondary growth. Secondary growth occurs because the stems of these plants also contain cylinders of dividing cells. As new cells are made, the stem increases in diameter. Secondary growth produces the tree rings that are sometimes used to age plants.

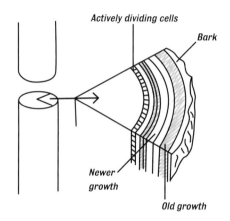

## Secondary growth in plants

*Plant stems grow thicker because of secondary growth.*

## Primary growth in plants

*Leaves, shoots and roots grow longer and larger because of primary growth.*

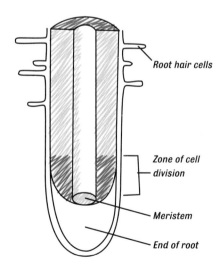

Just like animals, **plants** consist of a variety of cells, which are organised into **plant tissues**, organs and systems.

**Dermal** tissues are protective tissues that cover the surface of the plant. Just like our skin, they are the first line of defence against injury and infection.

**Vascular** tissues provide mechanical support and help to move materials through the plant. The two types of vascular tissue are xylem and phloem.

**Mesophyll** tissue is made up of different sorts of cells. It performs tasks like storage, photosynthesis, support and moving molecules around.

# Plant organs

The different plant tissues are arranged to form organs. Plants have three basic types of organ: leaves, stems and roots.

Most of the photosynthesis occurs in leaves. As well as receiving sunlight, leaves also control water loss, dissipate heat, provide protection, and exchange gases with the atmosphere.

Plants have evolved many different adaptations to suit their varied lifestyles. The spines of some cacti, such as the prickly pear, are actually leaves, as are the tendrils used by pea plants to cling on to their surroundings. Some leaves are covered in fine hairs, which help to keep the plant warm and prevent bugs from eating the leaves. Some emit strong-smelling molecules to deter herbivores, while others have evolved trap-like structures that help them to catch prey.

A stem is a plant organ containing leaves and buds. Its main job is to orient the shoot so it can receive maximum sunlight. Stems also help to transport fluids between the roots and the shoots, and elevate reproductive structures so that any pollen and fruit can disperse across a wider area.

## Plant organisation

*Compared with animal organs, plants organs are simple. They're just leaves, stems and roots.*

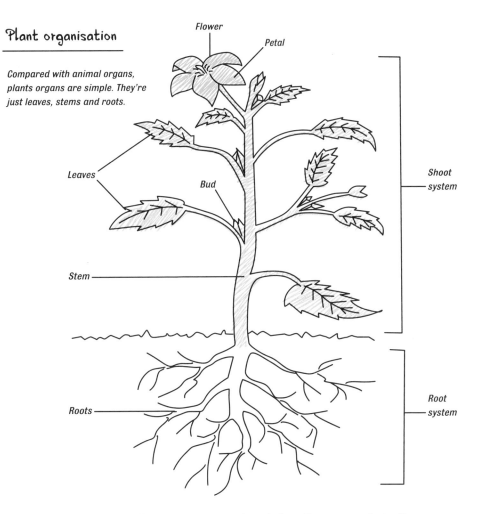

Tubers are actually stems. Potatoes, for example, are modified stems that are adapted for storing food. The stems or 'trunks' of the world's tallest plant, the California redwood tree, can grow over 100 metres (328 feet) tall.

The **root system** anchors the plant to the soil. This helps prevent it from being knocked down or eaten in its entirety. Crucially, roots absorb water and minerals from the surrounding soil. Some plants use their roots for communication. The roots of different trees are connected by a network of underground fungi, which help to relay messages between them. In arid landscapes, roots have to penetrate deep into the ground in order to find water. The shepherd tree, which grows in the Kalahari Desert, has roots that can extend down for more than 50 metres (164 feet).

# Plant systems

The cells, tissues and organs of a plant all work together to form systems. The leaves and the stems form the **shoot system**, and the roots form the root system. All of the structures are specially adapted for their various roles.

Leaves, for example, contain a variety of tissues organised so the leaf can work efficiently. The underside of the leaf is made of epidermal tissue. It is full of tiny pores called stomata.

## Transpiration

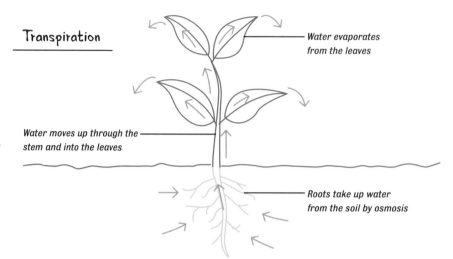

Water evaporates from the leaves

Water moves up through the stem and into the leaves

Roots take up water from the soil by osmosis

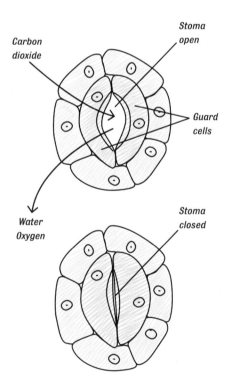

Carbon dioxide

Stoma open

Guard cells

Water Oxygen

Stoma closed

Plants need carbon dioxide for photosynthesis, so they take in carbon dioxide, and then release oxygen as a waste product. The gases enter and leave the plant via the stomata. When the stomata are open, carbon dioxide from the atmosphere diffuses into the tiny air spaces and then into the cells of the plant. At the same time, oxygen produced by photosynthesis, diffuses out to the atmosphere.

The stomata also help the plant to control how much water is lost from the leaf. The stomata are surrounded by pairs of guard cells, which control the size and opening of the pore. When the guard cells are swollen, the stomata

## Stomata

*Stomata open and close to help control the movement of substances into and out of the leaf.*

open. Water vapour evaporates from the cells that line the air space, and pass out of the leaf into the atmosphere via diffusion. This loss of water vapour is called transpiration.

As water evaporates from the leaves, more water is drawn into the roots and up the stem. This is called the transpiration stream. Transpiration is affected by environmental conditions. Humidity, temperature, wind and light intensity all affect how much water is lost from a plant.

Plants compensate for this by opening and closing their stomata. Most plants open their stomata during the day when photosynthesis is occurring, then close them at night when photosynthesis stops. If it is hot or dry during the day, the stomata may close early to avoid excess water loss.

Epidermal tissue covers the top surface of the leaf, but there are no stomata here. Instead, the tissue is specialised to prevent water loss. It is made from a single layer of tightly packed cells, and often has a waxy, waterproof outer coating.

In between the two surfaces, the mesophyll tissue is made of two layers of cells. The spongy layer contains cells that are loosely packed and surrounded by big pockets of air. This gives them a large surface area so gases can diffuse into and out of the cells. This is molecular transport on a small scale, from one cell to another.

Next to the spongy layer, the palisade layer is full of tightly packed cells that are full of chloroplasts. This is where most of the photosynthesis occurs.

Photosynthesis produces sugars, which then need to be transported from the cells where they are made to the rest of the plant. So, now the plant needs to transport molecules over much bigger distances. This is achieved by the plant's vascular system. The vascular system is a network of tubes that is hidden inside the leaves, shoots and roots. It can transport molecules all around the plant, from the tip of the roots to the tip of the shoots.

Phloem transports the sugars that are made in the leaves to the rest of the plant. This includes to the tips of the roots and shoots, where the actively dividing cells need energy to produce new cells for growth. The sugars are also transported to storage organs, like tubers, where they are laid down for future use.

Xylem carries water and mineral ions from the roots to the rest of the plant. This is important because the cells need these substances to help them build proteins and other molecules.

## Cross section of a leaf

*Leaves are specially adapted to maximise photosynthesis and the transport of substances into and out of the plant.*

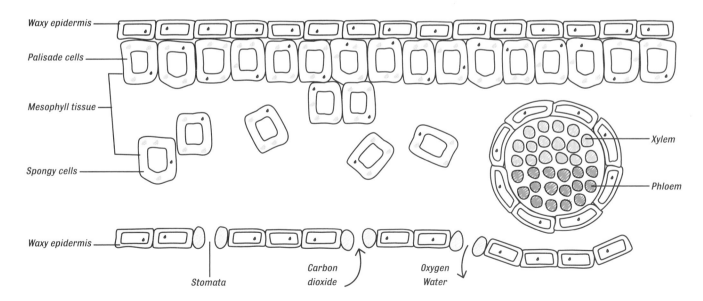

Waxy epidermis

Palisade cells

Mesophyll tissue

Spongy cells

Waxy epidermis

Xylem

Phloem

Stomata

Carbon dioxide

Oxygen
Water

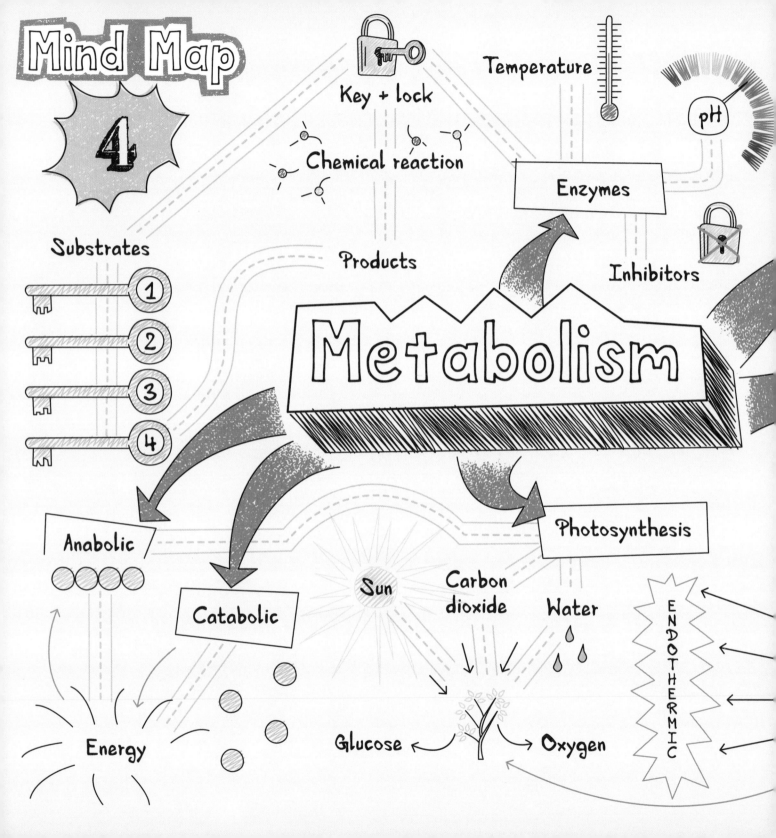

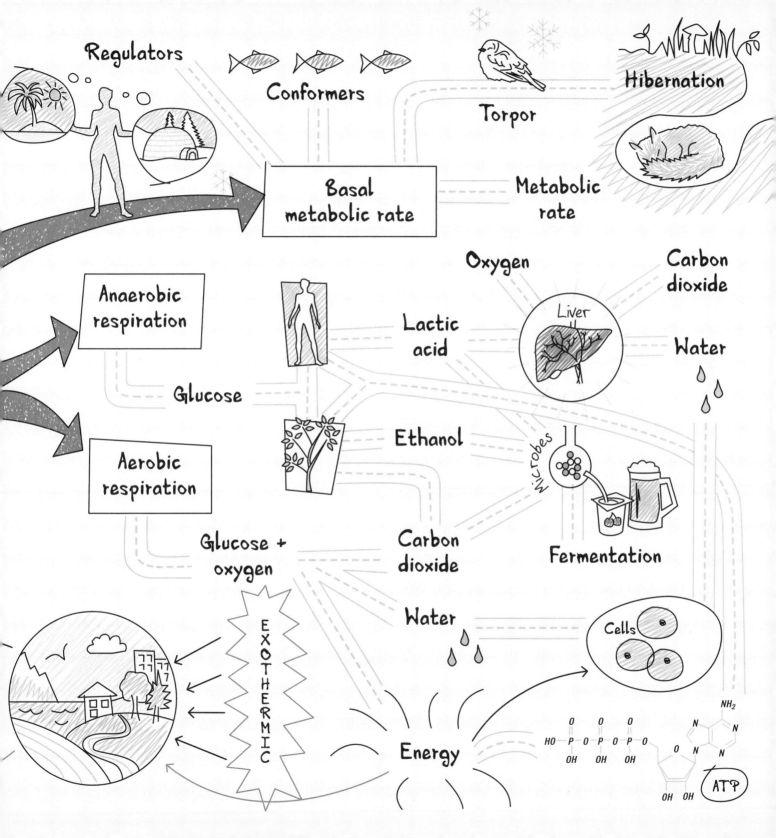

# Metabolic Pathways

While you are sitting still reading this book, it may not seem like your body is doing much, but if you were to peer inside any one of your trillions of **cells**, you'd find that it is a bustling hub of activity. The cells in our bodies are never at rest. Instead, they are busy carrying out all of the vital **chemical reactions** that are needed to keep us healthy and alive. These reactions are collectively referred to as an organism's **metabolism**.

Metabolism is absolutely vital to all living things. It helps organisms transform food into energy, break down waste products, and assemble the building blocks required to make cells, tissues, organs and systems. Metabolism is a process that begins the moment life starts, and continues all the way through our lives. If metabolism stops, then an organism will die.

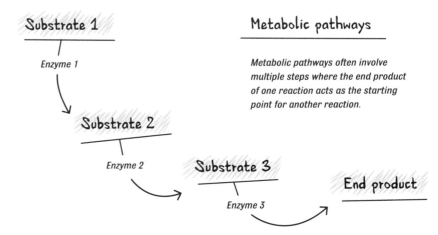

**Metabolic pathways**

*Metabolic pathways often involve multiple steps where the end product of one reaction acts as the starting point for another reaction.*

Often the chemical reactions are linked together to form chemical pathways. Each one begins with a specific molecule, which is then altered in a series of defined steps to form a useful end product. The molecules that are changed are called reactants or **substrates**, and a molecule called an enzyme controls each step in the pathway.

Metabolic pathways can be divided into two main categories. **Catabolic** pathways break down larger molecules into smaller ones. Along the way, **energy** is released. Respiration is a catabolic process.

**Anabolic** pathways build bigger molecules from smaller ones. These require energy. When cells build proteins from amino acids this is anabolic. When green plants combine **carbon dioxide** and **water** to make **glucose**, this is another anabolic process. Anabolic pathways are also called biosynthetic pathways, because they synthesise biologically useful products.

There are hundreds of different metabolic pathways. Most of them involve multiple steps and multiple enzymes. Imagine the subway map of a large, sprawling city. It will be

*Anabolism*

*Catabolism*

## Anabolism and catabolism

*Metabolic pathways fall into one of two categories: those that build up molecules, and those that break them down.*

similar to a map of the chemical pathways that occur inside your cells. Every line is a reaction and every station is a reactant or product.

If a particular enzyme or substrate is unavailable, then sometimes the end product can still be made using a different metabolic pathway. This is like taking an alternate route on the subway in order to reach your destination. It may take a little longer than the original path, but the end result is still the same.

Many metabolic pathways are shared between species. Glycolysis, for example, is the metabolic pathway that breaks down glucose to release energy. It is part of respiration. Nearly all living organisms carry out glycolysis as part of their metabolism. This suggests that the process has a deep-rooted evolutionary history, and that it persists to this day because it is so useful.

Other metabolic pathways are restricted to certain types of organism. Animals, for example, have specific metabolic pathways that enable them to produce vitamins and haemoglobin, and plants have specific metabolic pathways that help them to make chlorophyll and glucose.

An organism's metabolic pathways help to determine which substances will be toxic, and which will be tolerable. For example, some single-celled organisms use hydrogen sulphide as a nutrient, but the same gas is poisonous to animals. Bamboo contains the chemical arsenic. Arsenic can be toxic to mammals, yet giant pandas can eat bamboo with impunity. This is because they possess a unique way of metabolising arsenic, and they excrete it in their urine.

*Arsenic kills most mammals, but pandas can metabolise this dangerous chemical and eat bamboo that is laced with it.*

# Enzymes

Cells control their metabolic reactions via enzymes. Enzymes are molecules that accelerate chemical reactions. Each enzyme interacts with a particular substrate. They act as catalysts, increasing the speed of the reaction but not becoming permanently changed themselves.

Some household products use enzymes to speed up chemical reactions. For example, biological washing powders contain various enzymes that break down fats, proteins and other molecules that cause stains.

There are thousands of different enzymes. Most of them are proteins. These are large molecules that are folded up to form precise, distinct 3-D shapes. Their shape is intimately related to their function.

In order to work, the enzyme and the substrate need to bind with one another. This is like a **key** fitting into a **lock**. The substrate is the key and the enzyme is the lock. The enzyme's carefully crafted shape ensures that only the correct substrate will fit.

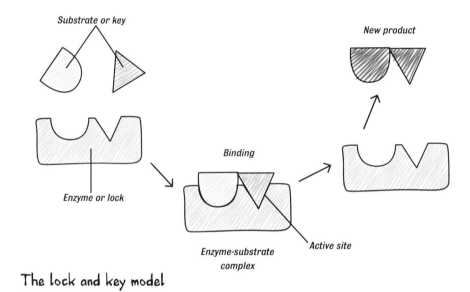

Substrate or key

Enzyme or lock

Binding

Enzyme-substrate complex

Active site

New product

## The lock and key model

*Substrates are like keys because they only fit certain enzymes or 'locks'. After the substrates bind to the enzyme, the reaction can proceed and the new product is made and released.*

The part of the enzyme where the substrate binds is called the active site. The active site will only bind to a particular substrate, so we say it is specific for that substrate. The active site also has a high affinity for its particular substrate. This helps to place the reactants into the correct position so that binding can occur.

Once the 'key' is in place, the enzyme and the substrate bind to one another to form the enzyme-substrate complex. This reduces the amount of energy that is needed to make the reaction happen – the activation energy of the reaction has been reduced.

With the substrate in place, the reaction takes place and the **products** are made. The products are not specific to the active site, so they are released from the surface of the enzyme.

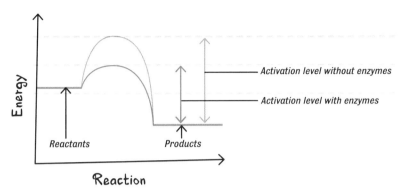

Activation energy

*Enzymes help reactions to run efficiently. If an enzyme is present, less energy is required to make a reaction happen.*

There are other, more complicated models of how enzymes work, but they are all based on this basic lock and key model.

The speed of metabolic reactions is influenced by a variety of factors. If the substrate is more concentrated, for example, there will be more 'keys' to fit the locks. The rate of the reaction increases until all of the active sites are occupied, and there are no more locks for the keys to fit. The enzyme is saturated. At this point, adding more of the substrate makes no difference to the reaction rate.

External factors such as **pH** and **temperature** also have an effect. Most of the reactions that occur inside our cells take place at relatively low temperatures. As the temperature increases, the rate of these reactions also increases.

Most human enzymes work at around 37°C (98.6°F), which is the internal temperature of the human body. Without enzymes to speed them up, the metabolic reactions that power the human body would occur too slowly to sustain life.

There is, however, an upper limit. Above 40°C (104°F), the enzymes start to break apart. This changes the precise 3-D shape so the substrate can no longer bind to the active site. The enzyme is unable to function as a catalyst and the reaction slows or stops. This is why it is dangerous to experience a prolonged, high fever. If your body temperature exceeds 41°C (105.8°F), your enzymes start to break down. This can be fatal.

Some enzymes buck this trend, and can work at much higher temperatures. For example, the bacteria that live in hot springs contain enzymes that can function at temperatures above 80°C (176°F). Similarly, while most enzymes work at a pH of 6 to 8, the enzymes in the stomach work well at a much lower, more acidic pH.

Body temperature and enzymes

*If the human temperature gets too high, enzymes break down and vital metabolic reactions are unable to occur. This can be life-threatening.*

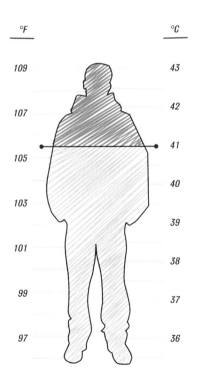

# Control of Metabolism

We have already seen how environmental factors, such as temperature and pH, can influence the metabolic reactions that occur in our cells. Molecules called **inhibitors** can also block the action of the enzymes that control these reactions.

There are three different types of inhibition.

1. Competitive inhibition occurs when an inhibitor binds to the active site of an enzyme and prevents the substrate from binding. Inhibitors can bind to enzymes because they are structurally similar to the enzyme's usual substrate. The effect can be reversed by increasing the concentration of the substrate, and flooding the enzyme's active sites with lots of substrate molecules. This dilutes the inhibitor so it stops working.

2. Non-competitive inhibition occurs when an inhibitor does not bind to the active site of the enzyme but does bind to another part of the enzyme. This changes the shape of the active site and prevents the usual substrate from binding. As a result, the rate of the reaction decreases. Unlike competitive inhibition, non-competitive inhibition cannot be reversed by increasing the concentration of the substrate.

3. Feedback inhibition occurs when the end product of a metabolic pathway binds to an enzyme at the start of the pathway, and shuts the pathway down. This is a form of negative feedback. It's a reversible process because when the concentration of the inhibiting molecule falls, the enzyme becomes able to catalyse the reaction and it starts all over again.

## Enzyme inhibitors

*Inhibitors block enzymes in various ways. Competitive inhibitors compete directly with the substrate while non-competitive inhibitors compete indirectly with the substrate.*

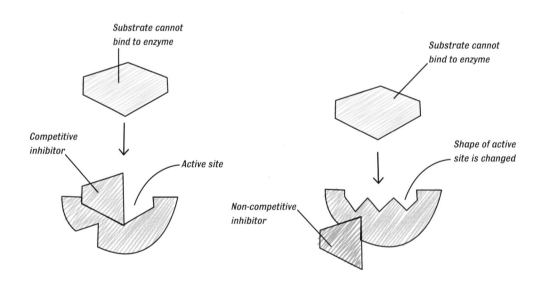

Substrate cannot bind to enzyme

Competitive inhibitor

Active site

Substrate cannot bind to enzyme

Shape of active site is changed

Non-competitive inhibitor

## Drugs and poisons

Enzyme inhibitors occur naturally in the body, but they are also used as drugs and poisons. Some antibiotics work by inhibiting key bacterial enzymes. For example, penicillin works by blocking the active site of an enzyme that is used by bacteria to build cell walls.

Toxins such as sarin, mercury and cyanide, are also enzyme inhibitors. In the mid-1990s, terrorists released the nerve gas sarin into the Tokyo Metro, killing a dozen people. The small molecule binds to and blocks a particular location on an enzyme called acetylcholinesterase, which plays an important role in the nervous system.

Most enzyme inhibitors are not at all harmful. They are a natural part of the cell's molecular makeup. If all of the cell's metabolic reactions were allowed to run at once, chaos would ensue. Inhibitors help to regulate this activity.

The body uses genetic mechanisms to help control the many metabolic pathways that exist. Organisms are able to determine when and where key enzymes are active. They do this by switching the genes that encode various enzymes and inhibitors on and off, and by regulating the activity of enzymes after they are made.

### Toxic inhibitors

*Some toxins are enzyme inhibitors. They cause ill health, and sometimes death, by disrupting key metabolic reactions.*

Metabolism is carefully controlled, so it's no surprise that the enzymes and their substrates are not distributed haphazardly through the cell. They can be found in precise locations. Some enzymes straddle the walls of cell membranes, while others float freely in the cytoplasm. Some of the enzymes that are involved in respiration are located inside the cell's mitochondria, while the enzymes that catalyse photosynthesis are found in plant cell chloroplasts.

Many enzymes need help from other molecules in order to work. These non-protein helpers are called cofactors. Cofactors are often vitamins, such as vitamin C, or metal ions of molecules like iron or zinc. The cofactors are either bound to the enzyme permanently, or they may attach and then detach along with the substrate.

In some cases, different enzymes are grouped together to form one big 'super-enzyme'. This is called a multi-enzyme complex. The arrangement of these enzymes influences the reaction that ensues. The product from the first enzyme becomes the substrate for the adjacent enzyme, and so on, until the final end product is released.

Scientists have identified thousands of different enzymes, in all sorts of different species. They have all arisen via evolution (see page 110). Enzymes are proteins and the instructions for making proteins are written in DNA. Random changes to the DNA of living things alter these instructions, so the enzymes that are made are slightly different too. They might bind different substrates, or assume some sort of new function. If the change helps the organism to survive, it is likely to persist.

# Respiration

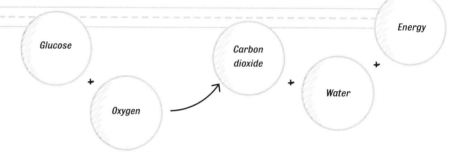

Glucose + Oxygen → Carbon dioxide + Water + Energy

Respiration is an important metabolic process. It occurs in all living cells, including plant and animal cells. Respiration is important because it allows living things to obtain energy from their food. Energy is needed for essential processes like growth, movement and reproduction.

Respiration is an enzyme-controlled chemical reaction. Aerobic respiration involves oxygen, but anaerobic respiration does not.

## Aerobic respiration

Aerobic respiration involves a complex series of chemical reactions and many different enzymes. During aerobic respiration, **glucose** reacts with **oxygen** to produce **carbon dioxide**, **water** and **energy**.

The raw materials for respiration come from the environment. We breathe in oxygen from the air, and obtain glucose from the food we eat. Carbon

dioxide is then released into the environment when we exhale, and the water that is made is used by our cells.

The energy that is made is in the form of a chemical called **ATP**. Most of this energy is used by the cells, but some is lost to the environment, making it slightly warmer. Aerobic respiration is called an **exothermic** reaction because it transfers energy to the environment.

Glycolysis is the first stage of aerobic respiration. It occurs in the cytoplasm of cells as glucose is broken down into smaller molecules. A small amount of energy is released.

The products from glycolysis then fuel the second stage. This is called the Krebs cycle. It occurs in the mitochondria, and it also generates a small amount of energy.

### How cells make energy when there is oxygen

*During aerobic respiration living things break down food to get the energy they need to live.*

The products from the Krebs cycle then fuel the third stage, which is called the electron transport chain. This also happens in the mitochondria. It generates a lot of energy.

The cells' mitochondria have a folded inner membrane (see pages 28–9). This provides a large surface area where enzymes and their substrates can bind, making respiration efficient.

# Anaerobic respiration

Most organisms respire aerobically, but some organisms and tissues can respire anaerobically when oxygen runs out. For example, our muscles respire anaerobically when we are exercising hard and our bodies cannot supply us with enough oxygen.

Like aerobic respiration, anaerobic respiration also breaks down glucose so that energy can be released in the form of ATP. Unlike aerobic respiration, the glucose molecules are not completely broken down. This means that less energy is produced. It makes anaerobic respiration less efficient than aerobic respiration.

Anaerobic respiration occurs in the cytoplasm of cells. In animal cells, the end products of anaerobic respiration are **lactic acid** and energy.

The buildup of lactic acid causes the burning sensation that occurs in your muscles when they are overworked. This causes muscle fatigue, and can prevent the muscles from contracting efficiently.

Excess lactic acid cannot be exhaled like carbon dioxide, so it has to be broken down by cells in the **liver**. This requires oxygen. Lactic acid and oxygen react together to produce carbon dioxide and water. The amount of oxygen needed to break down the lactic acid is known as the oxygen debt.

This explains why we sometimes continue to puff and pant after we have stopped exercising. It's the body's way of ensuring it gets enough oxygen to break down the lactic acid.

Plants and microorganisms can also respire anaerobically. Some **microbes** produce lactic acid when they respire without oxygen, others produce **ethanol** and carbon dioxide. Plants also produce ethanol and carbon dioxide when they respire anaerobically.

When microorganisms respire anaerobically, it is called **fermentation**. People have been harnessing the power of fermentation for thousands of years. This process is used to make certain foods like yogurt, sauerkraut and beer.

## Muscle fatigue

*Muscle fatigue occurs when lactic acid builds up in the muscles. The lactic acid needs to be broken down, but this takes oxygen.*

## How cells make energy when there is no oxygen

*Anaerobic respiration enables some living things to acquire energy from food in the absence of oxygen.*

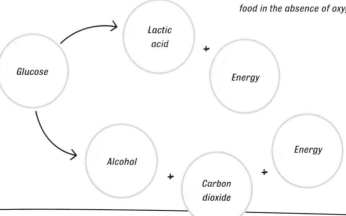

Glucose → Lactic acid + Energy

Glucose → Alcohol + Carbon dioxide + Energy

# Photosynthesis

Animals need to eat food in order to obtain energy. Plants do not. Instead, plants make their own food using photosynthesis.

Photosynthesis is a metabolic process. During photosynthesis, plants convert carbon dioxide and water into glucose and oxygen. Energy from the **Sun** is used to power this reaction.

Photosynthesis is an **endothermic** reaction because it requires energy from the environment to make it happen.

The process is, effectively, the opposite of respiration, which converts glucose and oxygen into carbon dioxide and water. The waste products of respiration serve as the starting point for photosynthesis, and the waste products of photosynthesis serve as the starting point for respiration.

The cells of plants and other photosynthetic organisms contain specialised organelles called chloroplasts (see pages 30–31). This is where photosynthesis occurs. Chloroplasts contain a specialised pigment called chlorophyll.

## How plants use sunlight

*Plants use sunlight to help them make food from carbon dioxide and water. This is called photosynthesis.*

Chlorophyll absorbs the light energy that has travelled about 150 million km (93 million miles) from the Sun. Because plants and algae need sunlight in order to photosynthesise, the process can only happen during the day.

Plants obtain their carbon dioxide from the air, and their water from the soil. The oxygen they produce is released into the atmosphere, which we then inhale when we breathe.

Some of the glucose produced by plants is used immediately. It is converted to energy via respiration. The rest is converted into useful substances such as starch and cellulose. Cellulose is used for strength. It is the main component of plant cell walls.

Starch provides an energy store that can be broken down by respiration at a later date. Starch is stored in the leaves, and sometimes in specialised tubers that grow underground. Potatoes and other root vegetables are packed full of starch, which humans also use as an energy source.

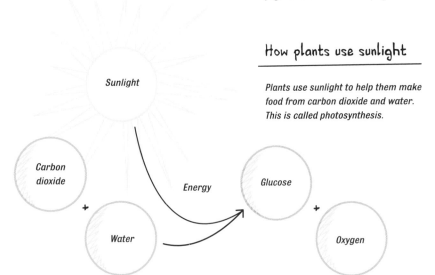

Sunlight

Carbon dioxide

+

Water

Energy

Glucose

+

Oxygen

## Photosynthesis in the field

*Plants need minerals to grow, as well as sunlight, water and carbon dioxide.*

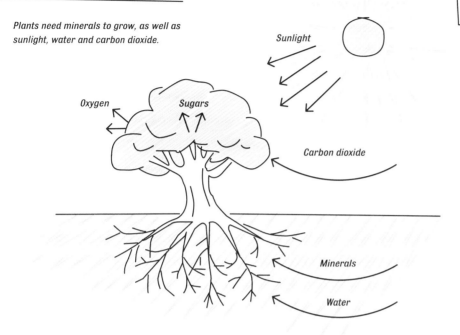

Deciduous plants, which lose their leaves in winter, make use of these energy stores to help them survive the winter. They provide food for the winter, and energy reserves to help fuel new growth in spring.

Plants also use some of the glucose from photosynthesis to make amino acids. They are formed by combining sugars with nitrate ions and other mineral ions from the soil. These amino acids are then joined together to form proteins, such as enzymes, which can be used by the plant. This process uses energy from respiration.

Most plants do not grow well without these minerals. When magnesium is lacking from the soil, plants struggle to make chlorophyll and their leaves turn yellow. When nitrates are missing, growth can become stunted.

Another use of the glucose from photosynthesis is to build up stores of fats and oils. These also act as energy stores. They are sometimes used to strengthen cell walls, or as an energy store inside seeds. Algae also photosynthesise. Some algae cells contain so much oil that scientists hope to harvest it as a possible biofuel.

## The vital role of photosynthesis

Plants make up most of the biomass on Earth. All life depends on photosynthesis, either directly or indirectly. When plants are consumed, the nutrients they manufacture become passed up the food chain.

Although each chloroplast is tiny, their collective action is immense. Photosynthesis produces an estimated 150,000 billion kg (165 billion tons) of carbohydrate per year. We use this for food, and have worked out how to boost the rate of photosynthesis by controlling the environment of the plants that we grow.

The first recorded greenhouse was built in 30 CE for the Roman Emperor Tiberius. Instead of glass, the structure was constructed from the mineral mica. It provided the emperor with cucumbers all year round.

Today, commercial greenhouses manipulate light, temperature, water, carbon dioxide and nutrients, so all sorts of plants can be grown all year round.

# Metabolic Rate

All organisms must break down fuel in order to keep their cells and tissues running. Metabolic rate is a measure of how quickly this process occurs. More specifically, an organism's **metabolic rate** is the sum of all the energy that is used in a given amount of time – usually per day. It is measured in joules (J), or in calories (cal) and kilocalories (kcal). One kilocalorie equals 1,000 calories or 4,184 joules.

Metabolic processes still happen when an organism is asleep or inactive. The body still needs energy to maintain vital processes such as respiration and keeping the heart pumping. This is called the **basal metabolic rate**. The basal metabolic rate is always lower than when an organism is being active.

Metabolic rate can be measured in different ways. Because respiration releases energy in the form of heat, metabolic rate can be measured by monitoring the amount of heat that an animal loses. Scientists use a device called a calorimeter. This is a sealed, insulated box with a thermometer to measure temperature. Another option is to measure the amount of oxygen that is consumed, or the amount of carbon dioxide that is produced by respiration.

## The relationship of metabolic rate to body size

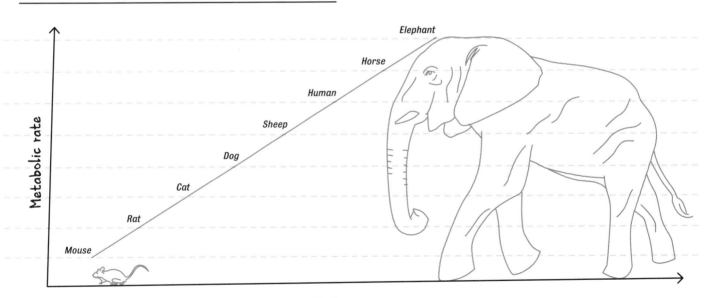

Metabolic rate

Mouse
Rat
Cat
Dog
Sheep
Human
Horse
Elephant

Body mass

Metabolic rates vary wildly between individuals and species. As a general rule, organisms that have more mass have a higher metabolic rate than organisms with less mass. This makes sense because larger animals require more energy to keep their cells and tissues working than smaller animals.

However, when metabolic rate per mass is studied, the situation is reversed. An ounce of mouse tissue, for example, requires about 20 times as many calories as an ounce of elephant tissue, even though the elephant consumes far more calories than the mouse. The smaller animal has a higher metabolic rate per ounce, so it needs a higher rate of oxygen delivery. We say that body size is inversely proportional to the energy needed to maintain each ounce of body mass.

Animals with high metabolic rates need to be able to deliver oxygen efficiently to their cells. Birds and mammals have higher metabolic rates than reptiles and amphibians, which in turn have higher metabolic rates than fish. Their circulatory systems reflect these different energy requirements.

Mammals and birds have a double circulatory system (see page 47), which helps to maintain their high metabolic rates. This includes a four-chambered heart that enables oxygenated and deoxygenated blood to be kept separate.

Amphibians and most reptiles have three-chambered hearts. They contain two atria and one ventricle, which helps to prevent oxygenated and deoxygenated blood from mixing too much. The amphibian heart is a balloon-like structure that is unable to exchange gases efficiently, so frogs have evolved the ability to breathe through their moist skin.

Fish have the simplest hearts of all vertebrates. They have one atrium and one ventricle. This is a single circulatory system.

An organism's metabolic rate is influenced by many factors, including activity, temperature, pH and salinity. Organisms can be categorised into two different groups.

**Conformers**, such as lizards and fish, are animals whose internal environment is heavily influenced by the external world.

**Regulators**, such as humans, are able to control their internal environment via their metabolism. This means they can live in a wider range of habitats, but this ability comes at a cost. Regulating the internal environment takes energy, so regulators tend to have higher metabolic rates.

Sometimes, animals encounter conditions that they find harsh. It may be too hot or cold, or there may be too little food. To save energy, some animals enter a state of **torpor** where they reduce their activity and metabolism. **Hibernation** is a form of torpor.

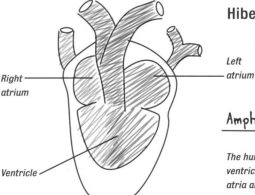

Right atrium

Left atrium

Ventricle

### Amphibian heart

*The human heart has two atria and two ventricles, but the amphibian heart has two atria and one ventricle. This is called an incomplete double circulatory system.*

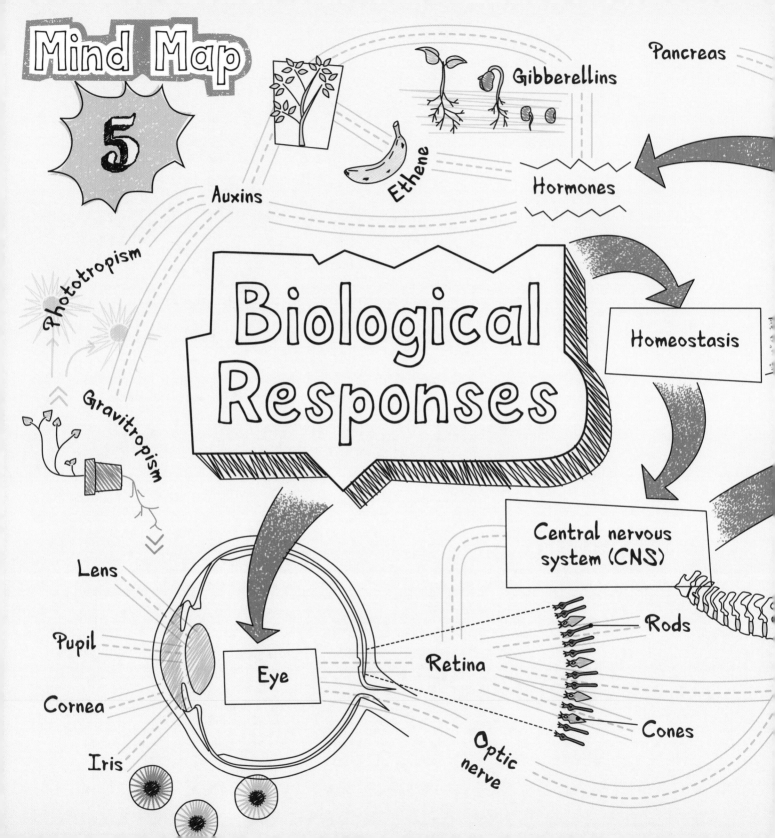

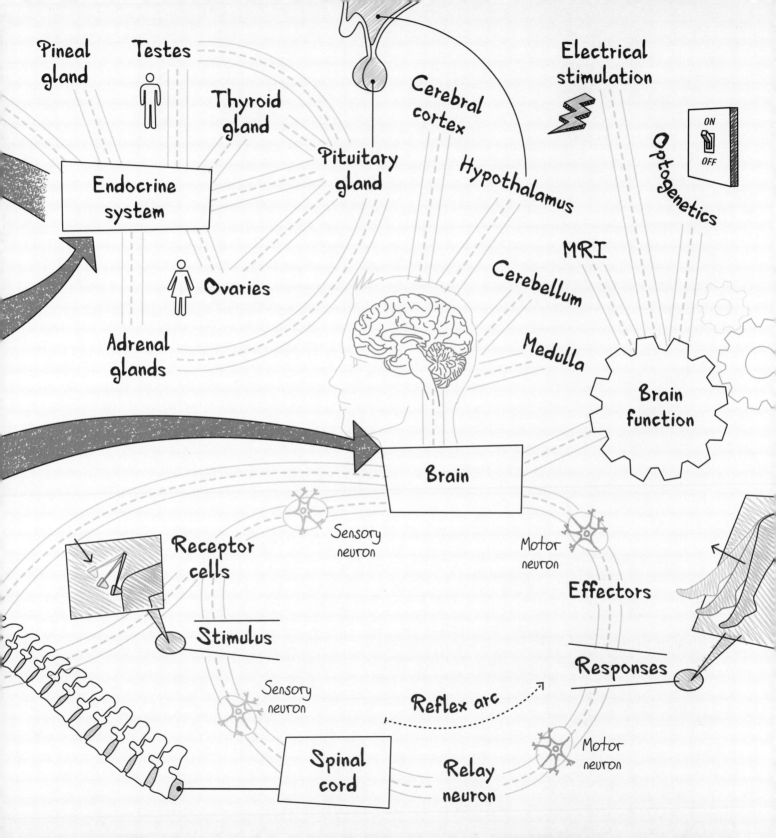

Pineal gland

Testes

Thyroid gland

Endocrine system

Ovaries

Adrenal glands

Pituitary gland

Cerebral cortex

Hypothalamus

Electrical stimulation

Optogenetics

ON
OFF

MRI

Cerebellum

Medulla

Brain function

Brain

Sensory neuron

Motor neuron

Effectors

Receptor cells

Stimulus

Responses

Sensory neuron

Reflex arc

Spinal cord

Relay neuron

Motor neuron

# Homeostasis in Animals

Human beings are an immensely successful species. People live just about everywhere – from the poles to the equator. We can cross deserts, climb mountains and trek through jungles, but how do our bodies cope?

In Chapter 4, Metabolism, we met the regulators. Regulators are animals that can adapt to their external environment by changing their metabolism and controlling their internal environment. This is an example of homeostasis.

**Homeostasis** is the ability of living things and cells to maintain a stable, relatively constant internal environment. Humans, for example, keep a fairly constant body temperature, blood pH and glucose concentration. It's a talent that gives regulators, like ourselves, the ability to exploit a wide range of habitats.

Homeostasis involves control and coordination. Think about heating a house. Most houses contain a heating system with a thermostat that can be preset to keep a constant temperature of, say, 20ºC (68ºF). When it's winter and the house is colder, the thermostat senses the drop in temperature and triggers the heating to come on. Conversely, in the summer when the temperature warms up, the thermostat will switch the heating off. The thermostat helps the house to achieve homeostasis.

## Homeostasis

*Our bodies contain an internal thermostat that controls body temperature in much the same way as the electrical thermostat in your house. This is an example of homeostasis.*

Many animals have similar control systems that work to keep variables, like temperature and pH, close to a set point. When the value fluctuates, the change or stimulus is detected by a sensor. These sensors can detect external change, such as fluctuations in temperature or light levels, and internal changes, such as water content and blood glucose levels.

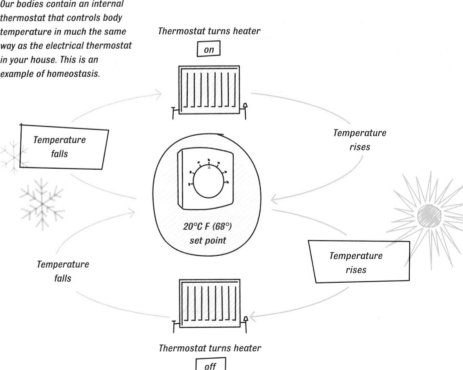

Thermostat turns heater [on]

Temperature falls

Temperature rises

20°C F (68°) set point

Temperature falls

Temperature rises

Thermostat turns heater [off]

The sensor sends a signal to a control centre, which then reacts by telling the body how to respond. The two major control centres are the nervous system, which includes the brain and spinal cord, and also the endocrine system. The tissues, muscles and glands that respond to these instructions are called **effectors**. Their **responses** help to restore conditions inside the body to optimum levels.

Homeostasis is important because the body needs a constant internal environment in order to operate effectively. The enzymes that control metabolic reactions, for example, work within a narrow temperature and pH range. If the values stray too far from the set point, the enzymes stop working.

Homeostasis is achieved using negative feedback. Negative feedback loops are control systems that dampen the original stimulus. Negative feedback loops play a major role in homeostasis in animals. The way our bodies control temperature is an example of negative feedback.

## Not too hot and not too cold

*The hypothalamus senses changes in temperature and responds by sending signals to the appropriate effectors.*

Stimulus $\rightarrow$ Sensor $\rightarrow$ Control centre $\rightarrow$ Effector

*During homeostasis, change is detected and responses are coordinated.*

If we get too hot, temperature sensors in the skin send information to the **hypothalamus**; the brain region that deals with temperature. The hypothalamus responds by sending nerve impulses to the sweat glands and blood vessels. These are the effectors. This triggers sweating, and as the moisture evaporates, it helps to cool the skin. The blood vessels that lead to the skin capillaries also become dilated. This allows more blood to flow through the skin, enabling more heat to be lost. As result, the body cools down and temperature is returned to normal.

If we get too cold, the hypothalamus reacts differently. This time it sends nerve impulses to the blood vessels, muscles and skin. The blood vessels that lead to the skin capillaries constrict. This reduces the blood flow through the skin, helping the body to conserve heat. Muscles contract rapidly, which causes shivering. This requires energy from respiration, and some of this energy is released as heat, helping to warm the body. The hairs on the skin stand up on end. This traps a layer of air above the skin, which helps to insulate the body and minimise heat loss. As a result, the body warms up and the temperature is returned to normal.

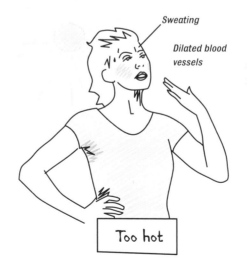

Sweating

Dilated blood vessels

Too hot

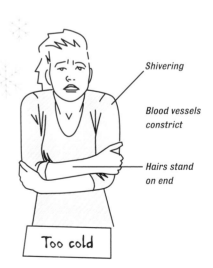

Shivering

Blood vessels constrict

Hairs stand on end

Too cold

# The Human Nervous System

The human nervous system is made up of the **central nervous system (CNS)**, which contains the brain and **spinal cord**, and the peripheral nervous system (PNS), which contains all of the nerves that relay information to and from the CNS.

Change is detected by specialised sensor cells called **receptor cells**. These cells are often clustered together in sense organs, such as the tongue, eyes and skin. Different receptor cells respond to different stimuli. For example, the receptor cells in the eye respond to light, the receptor cells in the ear respond to sound, and the receptor cells in the skin respond to touch and temperature.

Once a receptor cell has identified a **stimulus**, it relays this information to the CNS. The information is transmitted as an electrical impulse that travels along a neuron.

Neurons are bundled together in groups of hundreds or thousands. These are called nerves. Nerves are like electrical cables, and the neurons inside them are like the individual wires. The cells that carry information from your sense organs to the central nervous system are called **sensory neurons**.

The CNS receives and deals with an enormous amount of information. It coordinates the appropriate response, and then sends out messages along specialised nerve cells called **motor neurons**. Motor neurons carry electrical impulses from the CNS to the rest of the body. This enables the effectors to respond.

These electrical impulses zip around the body at speeds of up to 200 metres (656 feet) per second. This enables us to respond quickly to changes in the environment.

## Reflexes

Sometimes, we consciously decide how to react to a stimulus. If we're thirsty, we may decide to have a drink. If we're cold, we may decide to put on a sweater. Other times, we react quickly, without conscious thought. If we touch something sharp, for example, we automatically pull away. These quick, automatic responses are called reflexes.

### The nervous system

The human nervous system is made of two parts. The central nervous system contains the brain and spinal cord. The peripheral nervous system contains all of the nerves that run through the body.

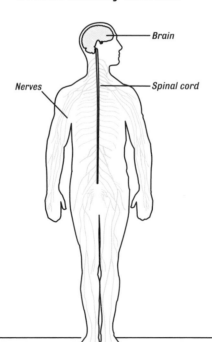

Brain

Spinal cord

Nerves

Reflexes are important because they help organisms respond quickly to danger. They also help living things to perform vital functions, such as breathing and digesting food. Life would be difficult if we had to consciously remember to breathe and break down food, but fortunately, our reflexes mean that we don't have to.

Simple reflexes, like withdrawing your hand from a hot object, involve three types of neuron: sensory neurons, motor neurons, and **relay neurons**. Relay neurons connect sensory neurons and motor neurons together. They are mostly found in the spinal cord.

Receptors in the skin of the hand detect the hot object, triggering an electrical impulse that travels along the sensory neuron. The impulse then passes along a relay neuron, and straight back along a motor neuron, all the way to the effectors. The effectors are the muscles

that contract in order to pull the hand away. This is called a **reflex arc**. Crucially, the conscious areas of the brain are bypassed, so the action happens automatically.

The various neurons do not connect with each other physically. Instead, they are separated by small gaps called synapses (see page 30). The electrical impulses traveling along the neurons cannot cross these gaps directly, so the electrical information is converted to chemical information. Chemicals called neurotransmitters are released from the neuron into the synapse. They diffuse across the gap and bind with the neuron on the other side. This generates a new electrical impulse that can then speed along its way. Synapses enable electrical signals to be passed between neurons.

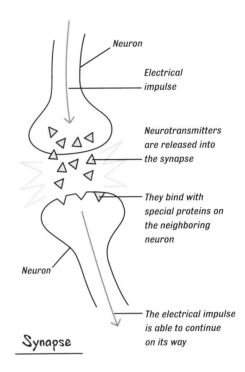

## Synapse

Synapses are the gaps between neurons.

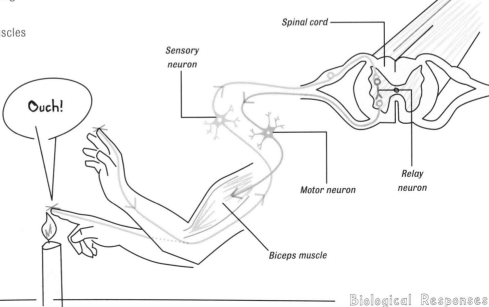

## Reflex arc

Neuronal pathways which bypass the brain help us respond quickly to potentially harmful conditions.

# The human brain

The adult human brain contains an estimated 100 billion neurons. Because a single neuron can connect with many other neurons via its synapses, the adult human brain contains trillions of different synapses. It is the most sophisticated computer on Earth.

The brain of an adult human weighs around 1.3 kg (3 lb). That's about the same as a chunky guinea pig. Fresh brains are soft and squishy, like blancmange. The brain is protected by membranes called meninges, and a sturdy layer of bone called the skull. Different regions of the brain perform different functions.

The outer layer, for example, is called the **cerebral cortex**. It is wrinkled and folded and covers much of the brain's surface. The cerebral cortex is involved in memory, attention, awareness, thought, language and consciousness. It is split into two halves or hemispheres. The left hemisphere receives information from, and controls the movement of, the right side of the body, and vice versa.

The **cerebellum** is toward the base of the brain. It helps control movement and balance. It receives sensory information about the position of different body parts. It doesn't initiate movement, but it does help with coordination.

Unconscious activities, such as breathing, heart rate and blood pressure, are controlled by the **medulla**. Specialised cells here also control sneezing and vomiting.

The hypothalamus is a small structure near the base of the brain that detects changes in blood temperature and water concentration. Another important function is that it links the nervous system to the endocrine system via the pituitary gland.

The pea-sized **pituitary gland** sits just below the hypothalamus. This vital structure secretes hormones.

## The brain

*The human brain is the most complicated machine on Earth. It contains billions of cells and many different interconnected regions.*

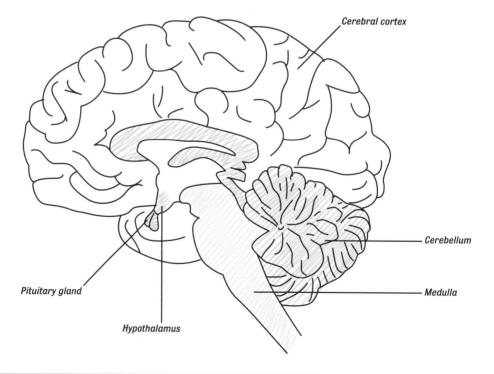

Cerebral cortex

Cerebellum

Medulla

Pituitary gland

Hypothalamus

## STUDYING BRAIN FUNCTION

Scientists still have much to learn about the brain. In the past, the only way to learn about what different parts of the brain did was to study people who had suffered brain damage.

Phineas Gage is the most famous example. Phineas was an American railroad worker. In 1848, he had a serious accident while laying a railway track. A large iron rod was driven right through his head. Remarkably, he survived, but his personality changed dramatically. He became impatient, lewd and profane. Doctors realised that the changes were caused by his brain damage. It helped them to realise that different parts of the brain have different functions, and suggested a key role for the brain in personality.

**Phineas Gage 1823–1860**

*Phineas's personality changed after an enormous iron rod was accidentally blasted through his skull.*

Today, we have sophisticated methods to measure brain function. Doctors routinely use MRI (Magnetic Resonance Imaging) scans to peer inside the brain. Strong magnetic fields and radio waves are used to produce detailed images of the brain and other body regions. Patients are asked to perform tasks while they are inside the scanner, like arithmetic or listening to music. By looking at the scan, scientists can tell which parts of the brain are involved.

**Electrical stimulation** is also used to study the brain. Electrodes are applied to the scalp, and then a weak electrical current is applied. If the visual area of the brain is stimulated, for example, the person may see flashes of light. If the area that controls movement is stimulated, the person may make an involuntary movement.

In recent years, researchers have acquired a new layer of sophistication. A technique called **optogenetics** uses genetics to make groups of neurons respond to light. It enables researchers to turn specific brain pathways on and off, just by turning light on and off. Now scientists are using it to discover many new things about the brain, and hope the technique will lead to new treatments for conditions such as epilepsy, Parkinson's disease and blindness.

# The eye

Many animals rely heavily on their sense of sight. The **eye** is an important sensory organ. Complex animals, like you and me, have evolved complex eyes, which bend and focus light rays in order to help us see.

The light rays enter the eye and fall on the **retina**. The retina is a layer of cells at the back of the eyeball. It's made up of millions of light-sensitive receptor cells, which respond to light by sending electrical impulses to the brain via the **optic nerve**.

The rest of eye has one job. It focuses the light on the retina. Light enters the eye via the **cornea**. The cornea is the outer transparent layer of the eye. It is curved so it bends the light slightly as it enters the eye.

The light then passes through the **pupil**. The pupil is actually a hole, surrounded by a muscular structure called the **iris**. The iris is made of muscles, which contract and relax to let in more or less light.

In dim light, the pupil becomes dilated to let more light enter the eye. In bright light, the pupil shrinks, reducing the amount of light that enters the eye. This is a reflex action.

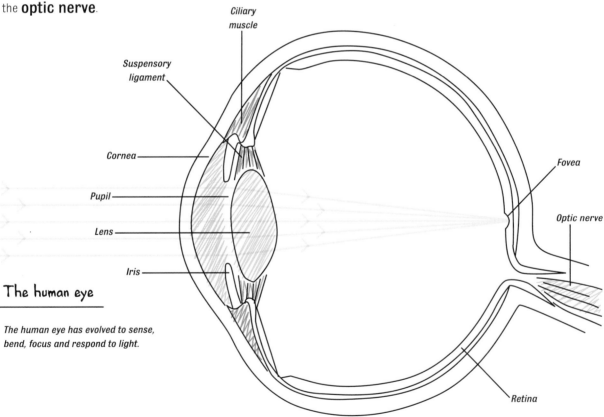

Ciliary muscle

Suspensory ligament

Cornea

Pupil

Lens

Iris

Fovea

Optic nerve

Retina

## The human eye

*The human eye has evolved to sense, bend, focus and respond to light.*

## The pupil adjusts to different light levels

*The pupil constricts in bright light and dilates in dim light.*

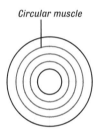

Circular muscle

Radial muscle

*Circular muscles contract in bright light*

*Radial muscles contract in dim light*

Just behind the pupil is the **lens**. This is where the fine focusing is done. The lens is a clear disc. It is held in place by the suspensory ligaments and ciliary muscles. When the muscles contract, the ligaments go slack and the lens gets fatter. This bends the light more as it passes through the lens. When the muscles are relaxed, the ligaments become taught and the lens becomes flatter and thinner. Now the light is bent less.

The ability of the lens to change the way the light rays are bent gives us the ability to change focus. It enables us to see clearly objects that are both close up and far away.

The human retina contains two main types of light-sensitive cell: rods and cones. Rods are more sensitive to light than cones, but they cannot distinguish between colours.

**Cones** provide colour vision. They work best in bright light. The human eye has three different types of cones: red, green and blue. They are sensitive to different parts of the visible light spectrum.

**Rods** are mostly found in the periphery of the retina and are used in peripheral vision. Cones are concentrated in a small, central area of the retina. The middle of this area is called the fovea. It only contains cones. When light falls on the fovea, the image that we see is really sharp.

Some animals have very different eyes. The compound eyes of flies, for example, are made up of hundreds of tiny, lens-capped structures. They produce hundreds of tiny, individual images that are then pieced together by the fly's brain.

## Compound eye of the fly

People who are colour-blind either lack certain cones, or have cones that are faulty. Other common eye problems include short-sightedness and long-sightedness. In both cases, the light does not focus on the retina, producing a blurry image.

Short-sightedness or myopia occurs when people cannot focus on distant objects properly. It can be caused by the eyeball being too long, or the lens being too thick and curved.

Long-sightedness occurs when people cannot focus on close objects. It can be caused by the eyeball being too short, or the lens becoming less elastic.

Lenses can often correct both conditions. Glasses and contact lenses bend the light in just the right way to compensate for the problems. Another option is laser eye surgery, which subtly reshapes the surface of the cornea.

# The endocrine system

The endocrine system is made up of glands that secrete chemical messengers called **hormones** into the bloodstream. The blood vessels carry the hormones around the body, and deliver them to target organs where they then have an effect.

Many processes in the body are controlled by hormones. These include growth and development, metabolism, sexual function, reproduction, sleep and mood.

The endocrine system is different to the nervous system, which uses electrical impulses to send signals via neurons. If the nervous system is like the internet traveling through wires, then the endocrine system is like the Internet using Wi-Fi. There are no wires or neurons. Instead, the hormones circulate through the body in the blood.

The body produces many different hormones. Compared with the nervous system, the effects of hormones are slower but longer lasting. Growth hormone is a slow-acting hormone. It is released by the pituitary gland in the brain and it stimulates the growth of bone and cartilage. Sex hormones, such as testosterone and oestrogen, are also slow acting.

*Adrenaline is released to prepare the body for fight or flight… or bungee jumping!*

There are, of course, exceptions. Sometimes hormonal responses are rapid. Adrenaline, for example, is produced in response to stressful or dangerous situations. Adrenaline quickly prepares the body for fight or flight by making the heart beat faster, increasing blood flow to the brain and muscles, and widening the pupils so that more light can enter the eye. The same effect occurs when we do something exhilarating. We call this an adrenaline rush.

## ENDOCRINE GLANDS

There are many endocrine glands dotted around the body. Many of these are regulated by a small, but powerful endocrine gland that is located in the brain – the pituitary gland.

The pituitary gland is also known as the master gland, because it influences other endocrine glands directly. It secretes many different **hormones**; some of which trigger other glands to release hormones. For example, the pituitary gland produces thyroid-stimulating hormone (TSH) that enters the blood and then acts on the **thyroid gland**. This stimulates the thyroid gland to produce various thyroid hormones, which help to regulate metabolism, appetite, and muscle function.

The pituitary gland also secretes hormones into the blood that have a direct effect on the body. Antidiuretic hormone (ADH) is one of these. It affects the amount of urine that is produced by the kidney.

Also, in the brain, the pineal gland produces the hormone, melatonin. Melatonin helps to control the body's sleep-wake cycle. During the day, the **pineal gland** is inactive, but when the sun goes down, it secretes melatonin into the blood. This leads to drowsiness. The blue light emitted by screens on mobile phones and TVs can dampen the production of melatonin. This is why many people advocate a short period of screen-free time before bed.

The **pancreas** is an endocrine gland (see page 44). As well as making digestive juices, it also secretes hormones such as insulin and glucagon. This helps to control the level of glucose in the blood.

The **adrenal glands**, which release adrenaline, sit just above the kidneys, while the thyroid gland, which releases thyroid hormones, is found in the neck.

In females, the **ovaries** release hormones such as oestrogen. Oestrogen controls the development of female secondary sexual characteristics, such as breasts and pubic hair, and is involved in the menstrual cycle and fertility.

In males, the **testes** release testosterone, which influences the development of male secondary sexual characteristics, such as facial hair and the Adam's apple, and is involved in the production of sperm.

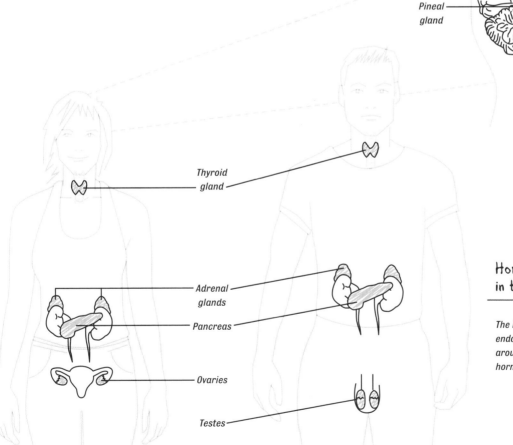

Pineal gland

Pituitary gland

Thyroid gland

Adrenal glands

Pancreas

Ovaries

Testes

### Hormonal regulation in the human body

*The human body contains various endocrine glands that are dotted around the body. They secrete hormones into the bloodstream.*

# Homeostasis in Plants

Just like animals, plants also need to be able to respond to changes in the environment. Unlike animals, plants don't have a nervous system, but they do produce hormones. Hormones made in one part of a plant are carried around the plant to other areas, where they have an effect.

Plants produce hormones without an endocrine system. Plants have no special hormone-producing glands. Instead, hormones are made in the plant's tissues.

Gravity and light are very important to plants. Plants need their roots to grow down, so they can absorb nutrients from the soil, and their shoots to grow up, so they can photosynthesise. How do they make this happen? They use hormones.

Plant hormones called **auxins** control growth in the tips of the plant's shoots and roots. Auxins have opposite effects in the shoots and roots. The hormones cause shoot cells to grow more, and root cells to grow less.

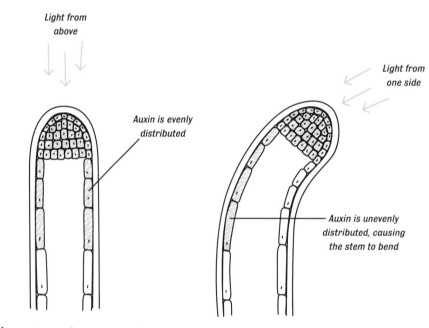

Light from above

Auxin is evenly distributed

Light from one side

Auxin is unevenly distributed, causing the stem to bend

## The action of auxin

*When light comes from one side, auxin diffuses away from the light. This causes cells to grow unevenly and makes the stem bend.*

Suppose a plant is directly lit from above. Auxin spreads evenly through the shoot tip, causing the tip to grow straight up.

Now suppose the plant is on a windowsill, and receives light from one side only. The auxin diffuses away from the light to the shady side of the stem. This makes the cells on the shady side elongate and grow more, so the plant bends toward the light. This helps the plant to maximise the amount of light available for photosynthesis. The response of a plant to light is called **phototropism**.

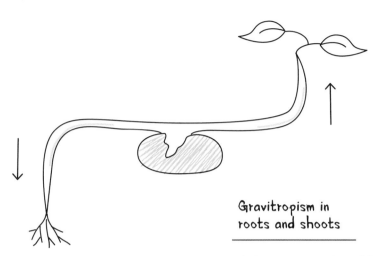

Gravitropism in
roots and shoots

*Auxins causes unequal growth in the roots and shoots of plants.*

In the roots, things are different. Roots do not respond to light, but they do respond to gravity. This is called **gravitropism**. When a seedling germinates, its roots need to grow down. Gravity causes auxins in the root to sink to the lower side of the root. Auxins, remember, cause roots to grow less, not more. So, the lower side of the root grows more slowly than the upper side, and the root curves down and into the ground.

Auxins are not the only plant hormones. **Gibberellins** are a group of plant hormones that are involved in growth and development. They are naturally produced by seeds, where they trigger germination. Low concentrations of gibberellins speed up germination. They cause cells to elongate, and plants to grow taller.

**Ethene** is another plant hormone. It's unusual because it is a gas. Ethene helps to control cell division as the plant is growing, and it helps fruits to ripen.

People have found ways to use plant hormones to help them grow plants more successfully. When gardeners take cuttings to generate new plants, they often dip the freshly cut shoots into rooting powder before planting them out. Rooting powder contains auxins. This helps to stimulate the growth of new roots, and helps the cutting to grow.

Perhaps paradoxically, auxins also make effective weed killers. Auxins stimulate plant growth. When the hormones are sprayed onto the leaves of weeds, the cells start to divide uncontrollably and die. Most weeds have broad leaves, while most crops have slender leaves.

When they are doused in weed killer, broad-leaved plants absorb more of the chemical than slender-leaved plants. This means that weeds die but the crops survive. Weed killers that target one type of plant over another are called selective herbicides.

Gibberellins are also used by plant growers. In the brewing industry, they speed up the germination of the barley seeds that are used to make malt. In the cut-flower industry, they are used to make plants such as geraniums and chrysanthemums flower year-round. In the food-growing industry, they are used to increase the size of seedless fruits such as grapes. Seeds produce gibberellins that help them to grow, but seedless varieties need a little extra help.

Ethene is used in the food industry to control the ripening of fruit. Bananas, for example, are often picked and transported while they are still green. When the journey is over, they are treated with ethene and then sent to the shops where they ripen.

If you keep ethene-ripened bananas in your fruit bowl alongside regular fruit, watch out! The ethene from the bananas will cause the untreated fruit to ripen.

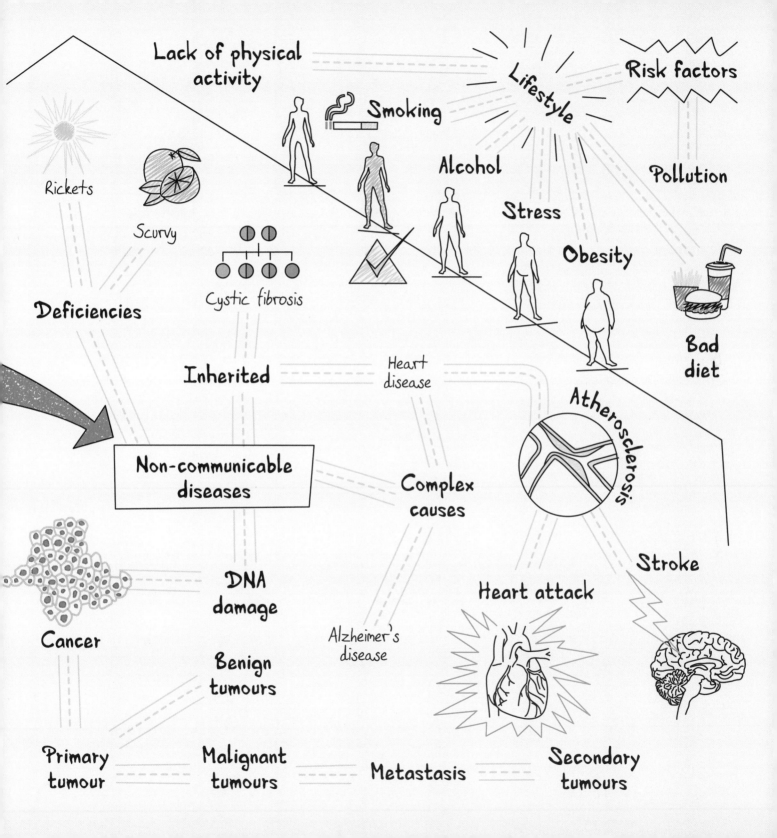

# Global Health and Disease

It is important to have a healthy lifestyle. Healthy eating, for example, can help to lower your risk of developing conditions such as diabetes and heart disease. People's life expectancy at birth is a good indicator of how well or otherwise they will be in later life. Over the past 25 years, life expectancy at birth has increased. The global average life expectancy of a person born in 1990 was 64 years. In 2013, it had increased to 71 years.

Children are always hardest hit by disease. Despite recent advances, every year around 6 million young children die from preventable causes. Most are in resource-poor countries. Children in sub-Saharan Africa, for example, are 15 times more likely to die before their fifth birthday than children in the developed world. Most of these deaths are preventable. They occur from a lack of basic resources, such as food and clean water, and from illnesses like malaria and measles.

These are **communicable diseases**, so called because they can spread from person to person. Communicable diseases affect adults too. Today, three of the top 10 causes of death are communicable diseases, and most occur in poor and developing countries.

**Non-communicable diseases** cannot be passed between people. Two-thirds of all deaths are caused by non-communicable diseases. They are the leading cause of disease around the world. **Heart disease** and **stroke**, for example, are the world's biggest killers.

As healthcare and living standards improve, and people live longer, the number of people living with ill health is set to increase. Deaths due to dementia have more than doubled over the past 20 years. **Alzheimer's disease** and other dementias are now responsible for the deaths of around 2 million people every year.

## Top 10 global causes of death

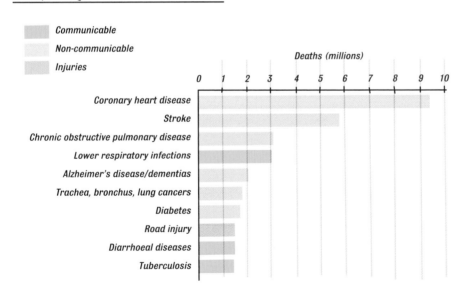

Legend:
- Communicable
- Non-communicable
- Injuries

Deaths (millions)

0 1 2 3 4 5 6 7 8 9 10

- Coronary heart disease
- Stroke
- Chronic obstructive pulmonary disease
- Lower respiratory infections
- Alzheimer's disease/dementias
- Trachea, bronchus, lung cancers
- Diabetes
- Road injury
- Diarrhoeal diseases
- Tuberculosis

# What makes us sick?

Communicable diseases are caused by microorganisms, including bacteria, viruses, fungi and protists. When they cause disease, these microorganisms are called **pathogens**. Non-communicable diseases, such as heart disease and **cancer**, are caused by a variety of different factors including genetics and **lifestyle**.

Both types of disease are major causes of ill health, but other factors also influence our health and well-being.

**Stress**, for example, is a normal part of everyday life, but researchers now realise that excessive amounts of stress can be damaging. Increased stress levels raise the risk of developing a wide range of health problems including heart disease and mental health problems, such as depression and anxiety.

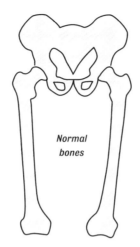

*Normal bones*

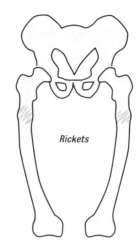

*Rickets*

### Rickets

*Rickets results in bendy, breakable bones. It's often caused by a lack of vitamin D.*

**A healthy diet** is important. It's not just about getting enough food, it's also about getting the right balance of nutrients. **Rickets**, for example, is a disease that affects bone development in children. People with rickets tend to have soft, breakable bones and bowed legs. Most cases of rickets are caused by a lack of vitamin D or calcium. Vitamin D comes from the Sun, but it's also found in some foods, such as oily fish and eggs. Calcium can be obtained from dairy products and green, leafy vegetables like broccoli and cabbage.

Other factors also play a role. Some diseases, for example, are more common in men than women, and vice versa. Ethnicity, financial status and the amount of available healthcare also have an effect, as do the part of the world that you live in and the cleanliness of your environment.

A recent study revealed that just one in 20 people have no health problems, and that a third of the world's population is currently enduring more than five different ailments. Multiple health problems are common because different diseases interact. One problem can trigger another, or make an existing problem worse.

# Communicable Diseases

Communicable diseases are either caused directly by a pathogen, or by a toxin that is made by a pathogen. Some communicable diseases, like the common cold and conjunctivitis, are relatively mild, while others, like influenza, Ebola and HIV/AIDS, can be deadly.

Sometimes, communicable diseases can spread from one species to another. These are called zoonoses. In 2002, a zoonotic disease called severe acute respiratory syndrome (**SARS**) broke out in southern China. The disease spread to other countries, including Hong Kong and Canada. More than 8,000 people were infected, and 774 people died. The disease began in bats, then spread to civet cats, then jumped to humans.

Bacteria and viruses cause most of the communicable diseases that affect animals. Fungi and viruses cause most of the communicable diseases that affect plants. These pathogens spread via a number of different routes.

Some are spread by **direct transmission**. The viruses that cause HIV/AIDS and hepatitis, for example, can enter the body when an infected person has sex with a non-infected person. Athlete's foot, which is caused by a fungus, can be caught directly, through skin-to-skin contact, and indirectly, when people touch contaminated objects, like towels and socks. Plant diseases are often

spread via this route. A tiny remnant from an infected plant can sometimes ruin an entire field of crops.

Some communicable diseases are spread by **indirect transmission**. The **air** spreads many pathogens, including bacteria, viruses and fungi. Droplet infection is common in humans. The virus that causes the common cold, for example, is spread when an infected person sneezes or coughs.

Other diseases are spread by **water**. The bacterium that causes cholera can be found in water contaminated by faeces from a person that has the infection. The bacterium can spread when the water is drunk or used to prepare food.

## The spread of disease

*Sometimes pathogens can spread from one species to another.*

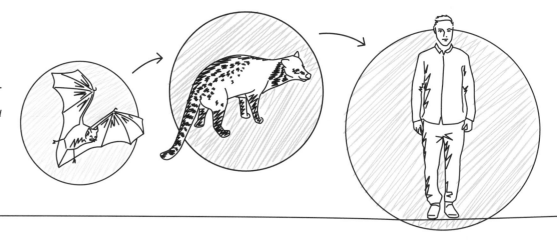

# Preventing infections

Today, the importance of handwashing and other simple **hygiene** measures is well known, but this has not always been the case. At the start of the nineteenth century, the causes of disease were unclear. Then along came Ignaz Semmelweis (1818–1865).

Semmelweis was a Hungarian doctor who worked on the maternity ward of an Austrian hospital. At the time, many new mothers died from 'childbed' fever shortly after giving birth, but Semmelweis noticed a pattern.

Mothers who were attended by doctors and medical students were more likely to die than those who were aided by midwives. Often the doctors and medical students went straight from dissecting corpses to delivering babies without washing their hands, so Semmelweis thought that they might be transferring the cause of the disease. He insisted the medics wash their hands, and shortly after, the number of deaths fell. It was the start of preventative medicine.

A short while later, French biologist Louis Pasteur built on this work, when his experiments proved that germs cause disease. He developed **vaccines** against diseases such as anthrax and rabies, and pioneered the germ-killing process that still bears his name: pasteurisation.

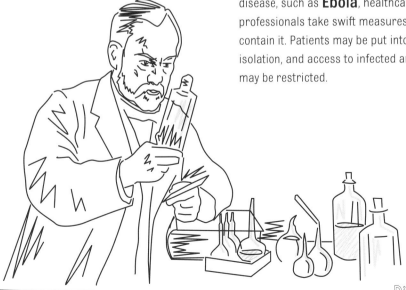

## Louis Pasteur
## 1822–1895

*The experiments of this famous French biologist proved that germs cause disease.*

## Now wash your hands

*Poor hygiene makes it easy for disease-causing germs to spread. This is why handwashing and other preventative measures are so important.*

Following in his footsteps, British surgeon Joseph Lister (1827–1912) pioneered the use of sterile surgery. He used antiseptics to destroy pathogens in the operating theatre. Little by little, the 'germ theory of disease' gained prominence, and, as our understanding of pathogens grew, lives were saved.

Now, sterile techniques, antimicrobial chemicals and vaccines remain at the front line of disease prevention. If there is a serious outbreak of an infectious disease, such as **Ebola**, healthcare professionals take swift measures to contain it. Patients may be put into isolation, and access to infected areas may be restricted.

# Bacterial diseases

Our bodies contain trillions of **bacteria**. Most of these do us no harm, and many are actually beneficial. Some gut bacteria, for example, help to protect the body from pathogens, but others cause disease.

*Salmonella* bacteria cause food poisoning. They live in the guts of various animals and can be found in undercooked foods such as meat, eggs and poultry. When they invade our bodies, they secrete toxins and disrupt the normal balance of gut bacteria. This causes cramps, vomiting and diarrhoea. The symptoms usually clear up on their own, but in countries where there is malnutrition, things can be more serious. The World Health Organisation (WHO) estimates that about 2 million people – mainly children under five – die from diarrhoeal diseases every year, including salmonella poisoning.

Other bacterial diseases include **tuberculosis**, which kills around 2 million people a year, mostly in sub-Saharan Africa, and **gonorrhoea**, which is transmitted sexually and can lead to infertility if left untreated. Plants also suffer from bacterial disease.

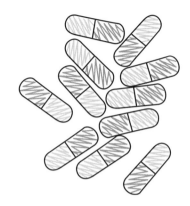

In the early twentieth century, infectious diseases caused more than 30% of all deaths in the United States. Now the figure is 10 times lower. Vaccinations, hygiene, and improved living standards have all played a part, but **antibiotics** have also played a major role.

Antibiotics, such as penicillin and amoxicillin, are molecules that kill bacteria. Some antibiotics target specific bacteria, while others kill a wide range. Since their discovery over 80 years ago, antibiotics have saved millions of lives, but now some bacteria are evolving **resistance** to them. This means that some antibiotics that used to work are becoming ineffective. The bacterium **MRSA**, for example, is resistant to several widely used antibiotics. We call it a 'superbug'. The emergence of antibiotic-resistant bacteria is a serious worry, so scientists are working hard to develop new, effective antibiotics.

## Salmonella

*The* Salmonella *bacterium is commonly found in the guts of warm-blooded animals. If people eat food that is contaminated with* Salmonella, *it can make them sick.*

# Viral diseases

Viruses (see page 21) can infect all types of organisms, including bacteria, plants and animals. The diseases they cause can be mild or serious. **Antibiotics** cannot cure viral diseases, and it's very difficult to make effective antiviral drugs. This is because viruses use the host's cells to replicate, making it hard to create a drug that kills the virus without also killing the host cell.

**HIV/AIDS** is a viral disease. It can be transmitted sexually and by the exchange of fluids like blood, which occurs when unscreened blood is used for transfusions and when drug users share needles. It is caused by the human immunodeficiency virus (HIV). Around 35 million people are thought to be infected. Infection causes mild, flu-like symptoms. When the symptoms pass, the virus remains hidden, weakening the host's immune system. This can lead to AIDS. When someone has AIDS, their immune system is so badly damaged that they cannot deal with infections and certain cancers.

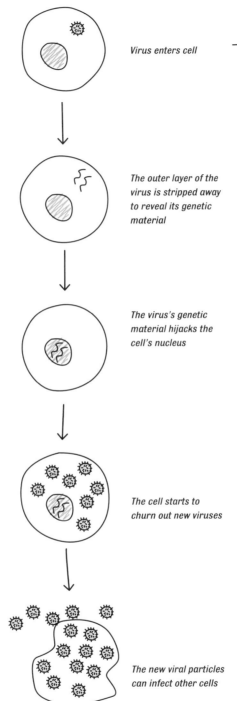

*Virus enters cell*

*The outer layer of the virus is stripped away to reveal its genetic material*

*The virus's genetic material hijacks the cell's nucleus*

*The cell starts to churn out new viruses*

*The new viral particles can infect other cells*

## Viral replication

*Viruses hijack the host's DNA and can churn out many infectious copies, which then infect other cells.*

There is no vaccine or cure for HIV/AIDS, but it can be prevented by practicing **safe sex**, avoiding needle-sharing and screening blood for transfusions. **Antiretroviral drugs** can help to keep AIDS at bay, and help HIV-infected people live a normal life. Unfortunately, most people with HIV live in sub-Saharan Africa where antiretroviral drugs are scarce. In these places, the life expectancy for someone with HIV/AIDS is very low.

Some viral diseases can be successfully prevented with vaccines. Vaccines are made from a dead or weakened version of a pathogen. This means the vaccine doesn't cause disease when it's injected, but does prompt the body to launch an immune response. White blood cells produce antibodies that destroy the vaccine molecules. Then, when the real pathogen is encountered, the antibodies set to work more quickly, destroying the pathogen before it can cause disease. **Measles**, for example, is entirely preventable by vaccination.

# Protist and fungal diseases

Protists cause a variety of different diseases in plants and animals. They are relatively rare, but the diseases they cause can be deadly. The disease-causing protist is often transferred to the host by a separate organism called a **vector**.

**Malaria** is the best-known protist disease. It is caused by the *Plasmodium* protist and spread by the *Anopheles* mosquitoes that harbour it. The **mosquito** is the vector for malaria.

The mosquitoes lay their eggs in water, which then hatch and transform into adults. Female mosquitoes need a blood meal in order to produce their eggs. When they bite a person, they transfer the *Plasmodium* parasites into the body. They travel in the blood to the liver,

## Malaria transmission

*The* Plasmodium *parasite that causes malaria has a two-stage life cycle involving two distinct hosts: humans and mosquitoes.*

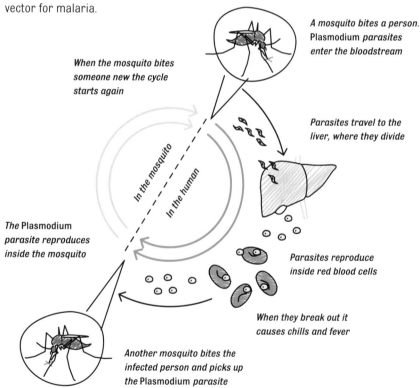

*When the mosquito bites someone new the cycle starts again*

*In the mosquito*

*In the human*

*A mosquito bites a person.* Plasmodium *parasites enter the bloodstream*

*Parasites travel to the liver, where they divide*

*The* Plasmodium *parasite reproduces inside the mosquito*

*Parasites reproduce inside red blood cells*

*When they break out it causes chills and fever*

*Another mosquito bites the infected person and picks up the* Plasmodium *parasite*

where they reproduce and make more parasites, which then fuel the illness. This causes fever, vomiting and diarrhoea.

Malaria claims around 400,000 lives per year, mostly in sub-Saharan Africa and mostly in children under five. It is entirely preventable using insecticides, bed nets and antimalarial **drugs**, and by purging the standing water where the parasite breeds. If it is caught early, malaria can be treated with a combination of drugs, but just as bacteria are becoming resistant to antibiotics, so too are the protists becoming resistant to antimalarial drugs.

Fungi can cause disease in people. **Athlete's foot**, for example, is minor, but some fungi cause problems that are more serious when they attack the lungs or brains of people who are already sick. Doctors treat **fungal** diseases with **antifungal drugs**. Fungal diseases are common in plants, where they represent a serious threat to agriculture. Two of the most serious fungal plant pathogens are the rice blast fungus and *Botrytis cinerea*. The rice blast fungus can devastate entire rice paddies, as well as fields of wheat, rye and barley. *Botrytis cinerea* is a ubiquitous fungus that causes grey mould. It infects fruit, vegetable and ornamental plants, so if you've noticed that your prepacked strawberries are starting to grow fuzz, this fungus is probably to blame.

# Non-Communicable Diseases

Non-communicable diseases are not caused by infectious agents, so they cannot be caught. They have a variety of different causes.

Some, like **cystic fibrosis**, are genetic, **inherited** disorders that people are born with. Others are caused by **DNA damage** that occurs during our lives. Most cancers fall into this category.

Another group is caused by **deficiencies**, such as a lack of essential vitamins or minerals. Anaemia, for example, can be caused by a lack of iron and vitamin B12 deficiency. **Scurvy** is caused by a lack of vitamin C.

Many non-communicable diseases have **complex causes** that are hard to pin down, and often genetic and environmental factors interact to cause the problems. Alzheimer's disease, many cancers, and diabetes fall into this category.

**Risk factors**, such as a person's age, lifestyle and environment, are known to influence the likelihood of developing certain non-communicable diseases.

Scientists sometimes spot patterns between non-communicable diseases, such as lung cancer and heart disease, and lifestyle factors, such as **smoking** and lack of physical activity. When two different factors or 'variables' seem to be related, we say there is a correlation. There is a correlation between the amount of cigarettes people smoke per year, and the number of deaths from lung cancer.

However, correlations do not prove that one thing causes another. Scientists did a lot of research to prove that smoking causes lung cancer. This included demonstrating that tobacco causes cancer in lab animals, identifying the cancer-causing chemicals in cigarette smoke, and understanding how these chemicals then lead to cancer. This last step is the causal mechanism.

A causal mechanism explains how one factor influences another via a biological process. If the causal mechanism for a disease is known, it helps researchers to design effective ways to prevent and treat it. If you are a smoker, for example, stopping smoking is the single, biggest thing you can do to reduce the risk of developing lung cancer.

## Cancer

Cancer is a common, non-communicable disease. In the United States, one in two women, and one in three men will develop cancer at some point in their lifetimes. Similar rates have been reported in other countries. Cancer is not a single disease. There are more than 200 different types of cancer. They can affect organs, like the lungs and oesophagus, tissues like epithelial and nervous tissue, and systems, such as the blood and immune system.

Cancer occurs when cells in the body begin to divide uncontrollably. This involves changes to the cell cycle. These changes are underpinned by genetic mechanisms. The rapidly dividing cells form a **primary tumour**.

In humans, cancer is not a communicable disease, but some non-human cancers are contagious. Tasmanian devils suffer from a contagious cancer called devil facial tumour disease. It is spread when the animals fight and bite one another, and has caused a significant drop in their numbers.

### Cancer cells

*Tumours form when cells divide uncontrollably.*

*Cancer cells growing through normal tissue*

Tumours can be benign or malignant. **Benign tumours** usually grow quite slowly. They tend to be contained in one place, usually within a membrane. They don't spread to other parts of the body, and are comprised of cells that are similar to normal cells. Often, they don't pose much of a problem, but if they grow too big, become painful or put pressure on a vital organ, they may require attention.

**Malignant tumours** are made up of cancer cells. They tend to grow faster than benign tumours, and can spread around the body. Malignant tumours are often called cancers. Sometimes, small parts of the cancer break away and spread around the body in the blood or lymphatic system. This is called **metastasis**. The cancerous cells end up in different parts of the body, where they can fuel the growth of more tumours. These are called **secondary tumours**.

### Tasmanian devil

*Tasmanian devils suffer from a contagious form of cancer that is transmitted when the animals bite one another.*

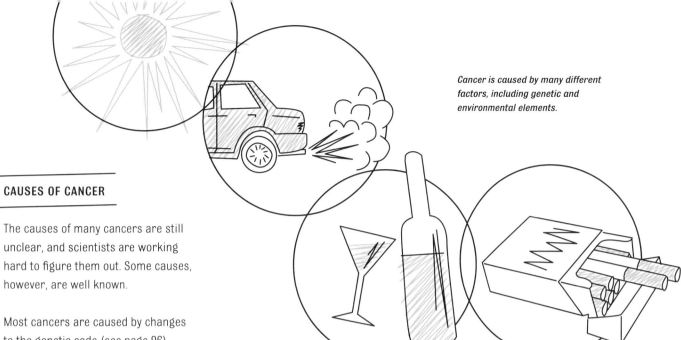

*Cancer is caused by many different factors, including genetic and environmental elements.*

## CAUSES OF CANCER

The causes of many cancers are still unclear, and scientists are working hard to figure them out. Some causes, however, are well known.

Most cancers are caused by changes to the genetic code (see page 96). Often, these changes occur during our lives, when we are exposed to DNA-damaging substances called carcinogens. **Alcohol** is a carcinogen, as are many of the chemicals found in tobacco smoke, such as tar and asbestos.

Some environmental pollutants are carcinogens. For example, air **pollution** has been shown to increase the risk of lung cancer.

Ionising radiation, such as UV light and X-rays, can also cause cancer. Melanomas occur when UV radiation from the Sun triggers DNA damage in the pigment-forming cells of the skin, causing them to divide uncontrollably.

Some cancers are caused by infections. Cervical cancer, for example, is usually caused by infection with the human papillomavirus (HPV). Teenagers are commonly vaccinated to prevent this disease.

Thanks to medical advances, many cancers that used to be incurable are now treatable, and the survival rates for cancer are improving all the time. There are many different treatments.

Sometimes, surgery is used to physically remove a tumour. Radiotherapy uses targeted doses of radiation to destroy the cancer cells, and chemotherapy uses anticancer drugs to either kill the cancer cells or make them self-destruct.

Some cancers respond to drugs, and some respond to hormone therapy. Hormone therapies block or lower the amount of hormones in the body (see page 76) to slow down or stop the growth of cancer cells.

If the cancer affects the blood or lymphatic systems, then bone marrow transplants may be an option. Bone marrow transplants contain stem cells. After they are transplanted, they can divide to form new red blood cells, white blood cells and platelets.

# Preventing disease and staying healthy

As research proceeds, new therapies for cancers and other non-communicable diseases are being developed. Treatments are improving, but we do not always have to know the exact cause of a disease in order to help prevent it. There are many things we can do to help ourselves live healthier lives.

Heart disease, for example, claims more than 9 million lives every year. The body becomes less able to deliver oxygen-rich blood to the heart as fatty deposits build up inside blood vessels. This is called **atherosclerosis**. This can cause angina, which is a tightening or pain of the chest. If a piece of the fatty deposit breaks away, it can cause a blood clot. If the blood clot blocks the supply of blood to the heart, it can cause a **heart attack**. When the blood clot blocks the blood supply to the brain, it can cause a stroke.

## Atherosclerosis

*Arteries become narrowed as fatty deposits build up inside them. This can cause angina, and lead to heart attacks and stroke.*

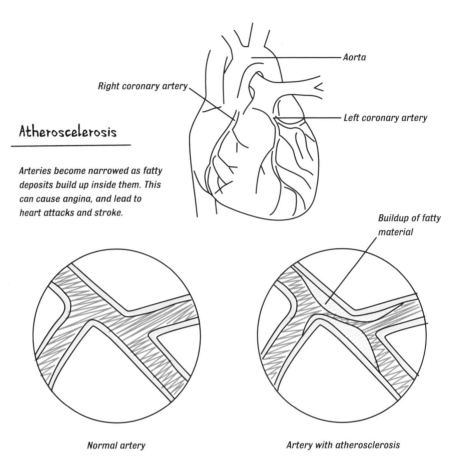

Aorta

Right coronary artery

Left coronary artery

Buildup of fatty material

*Normal artery*

*Artery with atherosclerosis*

There are many risk factors for heart disease. These include smoking, high blood pressure, high cholesterol and **obesity** (being overweight). Age and family history also play a role, as does alcohol consumption and stress. The more risk factors you have, the greater your risk of developing heart disease. The good news is that if you reduce these risk factors, you reduce the risk of developing heart disease.

The WHO estimates that 16 million people die prematurely every year from non-communicable diseases that could be prevented. Many of the risk factors for these diseases are things that we can change, so we can reduce the risk of developing non-communicable diseases.

Diet is particularly important. If you eat a **bad diet** and more food than you need, the excess becomes stored as fat. This can lead to becoming overweight and obese.

Body mass index (BMI) is a measure of obesity. Someone with a BMI of more than 30 is generally considered obese. A person with a BMI of more than 25 is said to be overweight.

In the past, obesity was considered a problem of high-income countries, but as living standards rise, it is now an increasing problem in low- and middle-income countries. In 2016, more than 1.9 billion adults were overweight. Of these, more than 650 million were obese.

Obesity and being overweight are risk factors for a number of diseases, including type 2 diabetes, heart disease and cancer. Today, being overweight and obese are linked to more deaths worldwide than being underweight.

This is almost entirely preventable. Keeping a healthy weight comes down to three things: eating healthier foods, eating less and doing more exercise.

## Lack of physical activity is

another major risk factor for disease. People who exercise regularly are less likely to develop heart disease than people who are physically inactive. They are also less likely to develop other diseases such as diabetes, Alzheimer's disease and some cancers.

People who exercise regularly have fitter hearts, bigger lungs, and more muscle than those who exercise less. Physical activity such as playing sports, going to the gym, and even walking regularly are good ways of keeping active.

The WHO advises that children under five should spend less time sitting and more time being active. Adults should do at least 150 minutes of moderate-intensity physical activity per week.

**Weight in kilograms**

Height in feet and inches

Weight in pounds

☐ Underweight

▨ Normal

▨ Overweight

▨ Obese

▨ Morbidly obese

## BMI calculator

*Body mass index is calculated by dividing a person's body mass by the square of their height. It is a measure of obesity.*

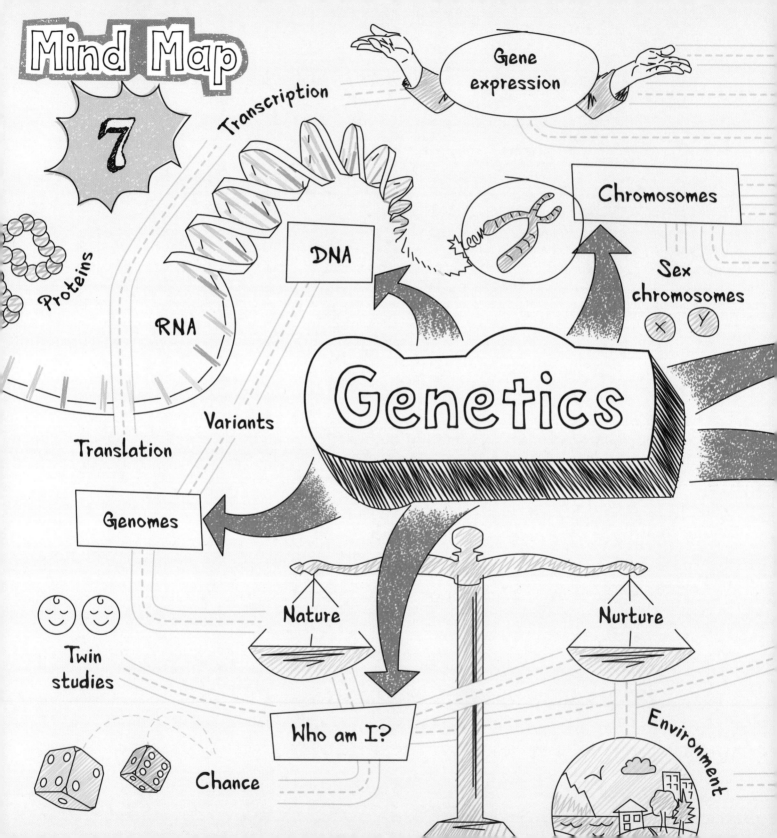

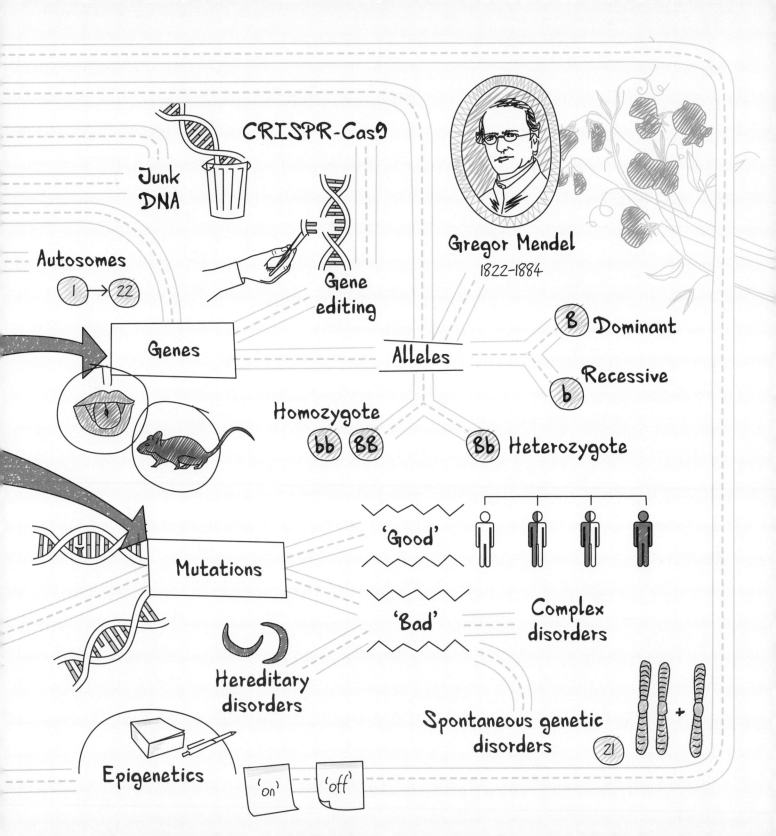

# The Molecule of Life

DNA is the molecule of life. It contains all of the instructions that a living organism needs to grow, reproduce and function.

**DNA** stands for deoxyribonucleic acid. It is a type of chemical known as a polymer, which consists of repeating subunits called nucleotides.

Each nucleotide is made up of three parts: a sugar, a small group of atoms called a phosphate group, and one of four different chemicals known as bases. The four different bases are adenine (A), cytosine (C), thymine (T) and guanine (G). All life on Earth is made from this four-letter alphabet.

In the 1950s, scientists discovered that DNA usually forms a double helix, like a twisted ladder. The sides of the ladder are chains of sugar and phosphate. The rungs are pairs of bases. The bases always pair together in a particular way. 'A' always pairs with 'T,' and 'C' always pairs with 'G'.

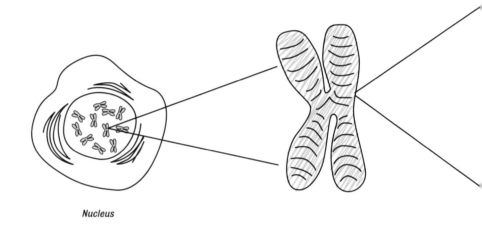

*Nucleus*

*Chromosome*

Almost every cell in the body contains DNA. It's a huge, long molecule that is tightly folded up so it can fit inside the cell's nucleus.

The genetic codes or **genomes** of living things are made up of billions of base pairs of DNA. The human genome, for example, is more than 3 billion base pairs long. The genome is the sum total of an organism's DNA.

The order of these letters is incredibly important. Imagine a recipe book. The order of the letters on the page tells the cook exactly what ingredients to use, and how to combine and cook them. Similarly, the letters in the genome tell our cells what they should be doing.

Inside the nucleus, the genome is broken up into manageable chunks called **chromosomes** (see page 34). These chromosomes contain important sequences of DNA called **genes**.

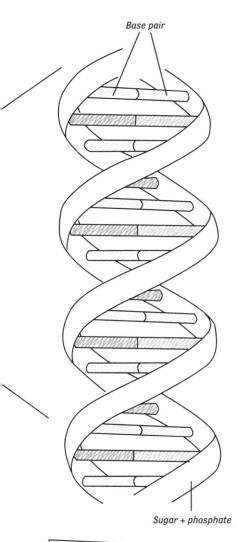

Base pair

Sugar + phosphate

**Bases**

| | |
|---|---|
| | Adenine |
| | Thymine |
| | Cytosine |
| | Guanine |

## Genes

Genes are specific sequences of DNA that code for **proteins**. The COL1A1 gene, for example, contains the instructions that tell a cell how to make collagen. Collagen is a protein that is found in many structures, including cartilage, bone, tendon and skin. Other genes code for the various enzymes that help our metabolism run smoothly (see page 57).

Each chromosome contains many different genes. The largest chromosome – chromosome 1 – contains about 8,000 genes. One of the smallest chromosomes – chromosome 21 – contains about 300 genes. Each gene is made up of hundreds or thousands of bases.

## Chromosomes and DNA

*DNA contains paired chemicals called bases, which are arranged into chromosomes, which reside in the cell's nucleus.*

The human genome was decoded over 15 years ago. Before it was finished, researchers set up a sweepstake to guess the number of genes in the genome. Their predictions were varied. Humans make more than 100,000 different proteins, so some thought there would be more than 100,000 genes. Others guessed much lower. The average guess was around 60,000. In the end, they were all wrong. We have since learned that the human genome actually contains around 20,000 genes.

Scientists are still learning a great deal from the human genome. It is helping them to pinpoint the causes of certain diseases, develop new therapies to combat them, and identify people whose genetic makeup makes them more likely to develop certain disorders. This is useful information, because it enables people to make positive lifestyle choices, such as quitting smoking or losing weight, in order to reduce the risk of developing disease (see page 93).

Researchers have now decoded the genomes of many other species, including rats, rice and rhinos. This is helping them to understand how different species are related, shedding light on evolution, and helping to fine-tune the tree of life (see page 13).

## GENE EXPRESSION

The human body is made up of hundreds of different specialised cell types (see pages 29–30). Each of these specialised cell types contains the same DNA, yet they look and act very differently. A neuron, for example, is very different to a muscle cell. Why is this?

Although the cells contain the same DNA and the same genes, not all of the genes are active. Different genes are switched on and off at different times. When a gene is switched on, we say that the gene is being expressed.

When a gene is expressed, the instructions encoded within it can be accessed by the cell and proteins can be made. So, the pattern of gene expression inside a neuron is different to the pattern of gene expression inside a muscle cell. Similarly, the pattern of gene expression inside a developing embryo is different to the pattern of gene expression inside an adult animal.

Although genes are important, most of the human genome is not made up of genes at all. Protein-coding sequences account for just 2% of our DNA. The remaining 98% is made up of non-coding sequences.

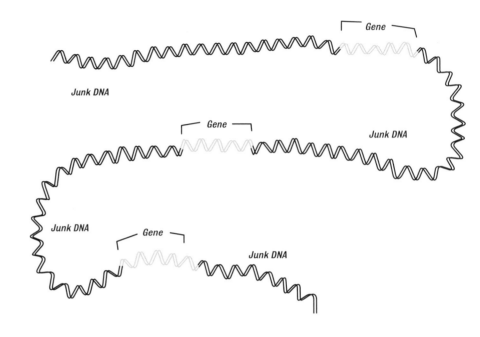

When geneticists first realised this, they referred to these 'non-gene' regions as **junk DNA**, because they thought that they had no useful function. Today, researchers are beginning to realise that many of these sequences do actually have a practical purpose.

Some of these non-coding regions are structural – they help to give the chromosomes their characteristic shape – whilst others are involved in gene expression. They act as switches to turn key genes on and off.

### Genes and junk

*Genomes contain a mix of protein-coding sequences, also known as genes, and non-coding sequences that are sometimes called junk DNA. As researchers learn more about it, they are beginning to realise that very often, junk DNA has a practical purpose.*

This can help to explain how the body is able to make so many different proteins with comparatively few genes. Genes can often control the production of many different proteins, and this, in turn, is dependent on gene expression.

# From genes to proteins

Genes contain the instructions for making proteins, but they do not make proteins directly. Other molecules are involved. This process involves two stages: transcription and translation.

**Transcription** occurs in the nucleus. If a gene is to be expressed, then its bases must first be copied or transcribed into **RNA** (see page 27). This creates a template that is a faithful copy of the original DNA sequence. The template is small enough to leave the nucleus through tiny holes in the nuclear membrane. It is called messenger RNA (mRNA) because it carries a genetic message from the DNA to the protein-making machinery of the cell. In the cytoplasm, the template attaches to the ribosome (see page 28).

There is now a change in language. The cell has to convert the sequences of nucleotides in the RNA into the amino acids that are needed to build a protein. This is called **translation**.

Each triplet of bases represents a particular amino acid (or a signal to stop making the protein). Carrier molecules in the cytoplasm are attached to specific amino acids. The carrier molecules

attach themselves to the template, so these amino acids can line up in exactly the right order. The carrier molecules keep bringing additional amino acids until the protein is complete. Then the protein detaches from the carrier molecules, and the carrier molecules detach from the template.

The fully formed protein now folds up into a complex 3-D shape that will enable it to carry out its functions in the cell. For example, if the protein is an enzyme, it will fold itself to form its characteristic active site (see page 56). If the protein is structural, like collagen, it will form fibrous structures. Now the protein is ready to do its job.

## Translation

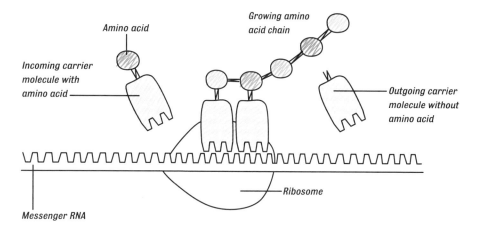

Amino acid
Growing amino acid chain
Incoming carrier molecule with amino acid
Outgoing carrier molecule without amino acid
Ribosome
Messenger RNA

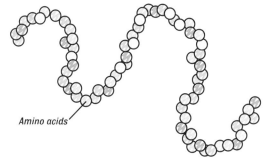

Amino acids

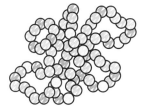

### Protein folding

*The freshly made protein is folded into a complicated 3-D shape.*

# Inheritance

Geneticists are interested in how genes, and the features that they influence, are passed on in families. Family members often look similar because they have certain genes in common. They may also share other less obvious characteristics, such as a predisposition to a particular disease. This is also due to shared genes.

We all have two copies of every gene and two copies of every chromosome (apart from the sex chromosomes, which we'll come to shortly). We inherit one copy from our mother, and the other from our father. They, in turn, received their genes from their parents, and so on back in time.

The copy that we receive from our mother may be slightly different to the copy we receive from our father. These different versions are called **alleles**.

Most features are the result of multiple genes working together. For example, height, personality and intelligence, are all influenced by many different genes.

Single genes, however, determine some characteristics. This includes fur colour in mice, and tongue-rolling and red-green colour blindness in humans. Some inherited diseases are also caused by single genes, such as cystic fibrosis and muscular dystrophy. These single-gene features tend to have clear-cut patterns of inheritance.

**Gregor Mendel** was born in Silesia, part of the Austrian Empire (now the Czech Republic). He was a monk and a botanist. It was Mendel who first described these patterns. He studied pea plants and painstakingly transferred pollen between plants to control their reproduction.

He found that when he bred purple-flowered plants with white-flowered plants, they always produced purple-flowered offspring. Then, when he bred these purple offspring with each other, three-quarters of their offspring had purple flowers, and a quarter had white flowers.

Mendel found that flower colour was determined by 'units of inheritance', which were passed from parent to offspring. We now realise that these units are genes, and that the pea plants were inheriting different alleles of the gene that controls flower colour.

Alleles can be **dominant** or **recessive**. A dominant allele is always expressed, even if there is only one copy present. When geneticists write about dominant alleles, they use a capital letter. So, 'P' can be used to represent the purple pea flower allele. Other examples of dominant alleles include the allele for black fur in mice, and the allele for wet earwax in people.

Recessive alleles are only expressed when an individual carries two copies, and so does not have a dominant allele of the same gene. Geneticists refer to recessive alleles with lowercase letters. So, 'p' can be used to represent the white pea flower allele. Other examples of recessive alleles include the allele for brown fur in mice, and the allele for dry earwax in people.

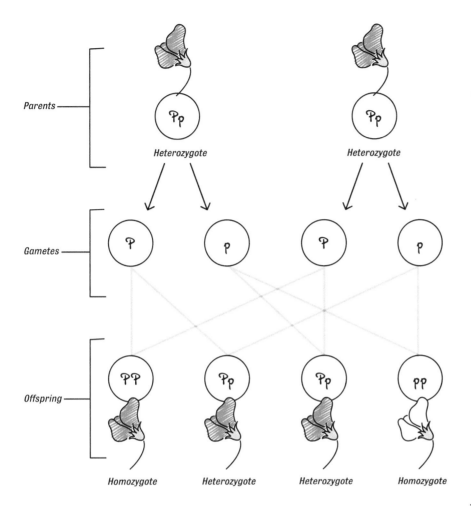

**Parents**

Heterozygote — Pp

Heterozygote — Pp

**Gametes**

P  p  P  p

**Offspring**

PP — Homozygote

Pp — Heterozygote

Pp — Heterozygote

pp — Homozygote

## The outcome of Mendel's pea experiments

*When Mendel crossed his first generation purple peas together, three-quarters of the offspring were purple, and one-quarter were white. 'P' is the dominant allele and 'p' is the recessive allele.*

If a single gene determines a feature – such as earwax or pea flower colour – then a Punnett square can be used to predict the outcome of crossing two different individuals together.

A Punnett square displays the alleles of a feature that the parents have, the possible gametes that can be formed from these, and combinations of alleles that could occur in future offspring.

If an organism contains two identical alleles of a particular gene – for example, two dominant alleles (PP) or two recessive alleles (pp) – then it is called a **homozygote**.

If an organism contains two different alleles of a particular gene – one dominant and one recessive (Pp) – then it is called a **heterozygote**.

In Mendel's experiment, the purple allele (P) was dominant, and the white allele (p) was recessive. The first-generation pea plants, which were all purple, were heterozygotes, and the second-generation pea plants, which were a mix of colours, were a mix of heterozygotes and homozygotes.

## Punnett square

*What happens when a brown homozygous and a black homozygous mouse are crossed?*
*B = black fur allele*
*b = brown fur allele*

| Gametes | B | B |
|---------|-----|-----|
| b | Bb | Bb |
| b | Bb | Bb |

*All offspring have black fur.*

# Sex determination

A single pair of chromosomes determines our biological sex. These are called the **sex chromosomes**.

Humans have 23 pairs of chromosomes. Twenty-two of these pairs are **autosomes**. Autosomes are any of the chromosomes that are not sex chromosomes. Each autosome in a pair is roughly the same size and shape. Autosomes contain many genes. These carry information about the many different features of your body.

The 23rd pair is the sex chromosomes. One is the male sex chromosome, and one is the female sex chromosome. The two sex chromosomes are usually very different in size.

Human females have two X chromosomes. Human males have one Y chromosome and one X chromosome. The Y chromosome is very small. Most of the genes that it carries are related to sexual characteristics. For example, the SRY gene triggers testis development.

Cells divide by meiosis in order to produce gametes, such as sperm and egg cells (see page 35). When this

## Human gamete production

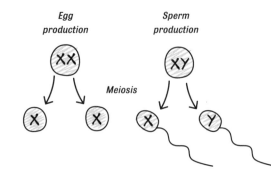

happens, one sex chromosome goes into each gamete. This means that human egg cells always contain a single X chromosome. Half of the sperm cells also contain an X chromosome, but the other half contain a Y chromosome.

Different species often have different numbers of chromosomes, and different sex chromosomes. This is due to evolution (see page 110). Birds tend to have 40 pairs of chromosomes. Instead of X and Y sex chromosomes, they have Z and W sex chromosomes. Here, females are the heterozygotes because they contain a Z and a W sex chromosome, and males are the homozygotes because their cells have two Z chromosomes.

## Chromosomes

*Most human chromosomes are autosomes but one pair, containing the X and Y chromosome, are responsible for determining sex.*

1  2  3  4  5

6  7  8  9  10  11  12

13  14  15  16  17  18

19  20  21  22

X  Y

— Autosomes —

— Sex chromosomes —

# Genetic disorders

Changes to the genetic code are called **mutations**. Mutations and mutants conjure up images of monsters and sci-fi movies, when in reality, mutations are a normal part of life on Earth. They can be **'good'** and **'bad'** for our health, and are an important part of evolution.

Some mutations crop up spontaneously. They may be the result of an error during cell division, or caused by environmental factors, such as pollution or smoking. If a mutation is present in a gamete, or a cell that gives rise to a gamete, then the mutation can be passed from parent to child. If the mutation has an adverse effect on the person, then the condition it causes is called a hereditary disorder.

---

### HEREDITARY DISORDERS

Sickle cell anaemia is a hereditary disease. It is caused by a mutation in a single gene. Specifically, one pair of bases is substituted for another. This one tiny change has profound effects.

## Red blood cell

*Normal cell*      *Sickle cell*

## The genetics of disease

*When a child inherits two copies of the sickle cell allele, it will develop the disease. When a child inherits one copy, it will be a carrier.*

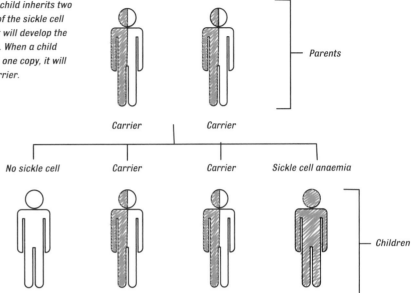

The gene codes for haemoglobin, which is the protein that ferries oxygen around the body. In sickle cell anaemia, the haemoglobin protein is abnormally clumpy, and the red blood cells that contain it have a rigid, sickle shape. They struggle to transport oxygen around the body, and can clog up the blood vessels. This can lead to problems such as stroke.

If an individual carries one copy of the sickle allele, they will not have the disease, but they will be a carrier. If their partner is also a carrier, then any children will have a one in four chance of inheriting the illness. This is because sickle cell anaemia is an autosomal recessive disease. To be affected, a person must inherit two copies of the problem allele.

Huntington disease is another single gene disorder. This devastating neurodegenerative disorder tends to affect people in their prime, who are around 30 to 40 years of age. People with this disease have an error in the huntingtin gene. Three nucleotides within the gene are repeated an abnormal number of times. This causes the brain to make an abnormal protein that damages and ultimately kills brain cells.

## SPONTANEOUS GENETIC DISORDERS

Down syndrome is an example of a genetic condition that can crop up spontaneously. People with Down syndrome have an extra copy of chromosome 21. This is a big mutation. It can occur when there is an error during meiosis, and the chromosomes in the parent cell fail to separate evenly.

Most people with Down syndrome live happy and fulfilling lives. They have some degree of learning disability, but this is different for each person. Sometimes people with Down syndrome have heart conditions, and problems with hearing and vision.

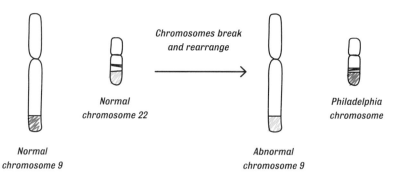

Normal chromosome 9

Normal chromosome 22

Chromosomes break and rearrange

Abnormal chromosome 9

Philadelphia chromosome

### Down syndrome

*People with Down syndrome have three copies of chromosome 21.*

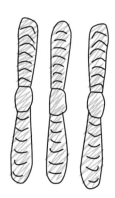

Chronic myeloid leukaemia (CML) is another genetic disorder that can occur spontaneously. CML is a type of blood cancer. Instead of having an extra chromosome, like in Down syndrome, people with CML have one very abnormal chromosome. It is called the Philadelphia chromosome, and it's formed when two chromosomes break apart and swap pieces. The result is an abnormal enzyme, which stimulates the leukaemia cells to keep dividing.

## COMPLEX DISORDERS

Many diseases have a genetic component, but very few of them are caused by mutations in a single gene. Instead, the roots of these complex disorders are far more complicated. They are caused by multiple mutations in multiple genes, and environmental and lifestyle factors. Heart disease, Alzheimer's disease, type 2 diabetes and most cancers fall into this category (see page 89).

### Philadelphia chromosome

*The Philadelphia chromosome contains an abnormal hybrid gene that causes white blood cells to keep on dividing.*

Although these disorders can run in families, they don't obey the classic Mendelian rules of inheritance. This makes it difficult to determine a person's risk of inheriting or passing on these disorders.

For example, researchers have identified some mutations that are more common in people with Alzheimer's disease. This is complicated, because not everyone with Alzheimer's carries the mutation. If you do have the mutation, you are more likely to develop Alzheimer's disease than someone who does not have it. However, exercise, eating well and staying healthy can still help to reduce the risk of developing Alzheimer's disease.

# Gene editing

Genetic disorders, such as cystic fibrosis and Alzheimer's disease, cannot be cured. Sometimes the symptoms can be eased, but the underlying genetic cause remains.

Gene therapy, currently in development, offers a way to fix the root cause of these disorders. Gene therapy seeks to replace faulty genes with healthy ones.

Scientists are currently very excited about a method called **CRISPR-Cas9**. CRISPR-Cas9 is a molecular tool that enables researchers to edit the genomes of living things with pinpoint precision. Bases can be added in, removed or altered at will. It's cheaper, easier and more accurate than previous gene editing methods.

In the last few years, the method has shown promise in various animal models of disease. For example, it has been used to restore vision in rats with a genetic form of blindness, and to boost levels of a missing protein in dogs with the genetic muscle wasting disorder, Duchenne muscular dystrophy.

Companies are now developing CRISPR-based treatments for a range of diseases. CRISPR-Cas9 has the potential, not just to cure certain genetic disorders, but also to prevent them from being passed down the generations.

In 2015, Chinese scientists showed that human gene editing is possible when they used CRISPR-Cas9 to modify genes in non-viable human embryos left over from in vitro fertilisation (IVF).

In the future, CRISPR-Cas9 could rid families of the inherited diseases that blight them, but some people are uncomfortable with this. They see it as playing God, and worry that scientists will be tempted, not just to cure disease, but to make other changes too. What if scientists were able to make 'designer babies' engineered to be smarter, stronger or more resistant to disease? Now society as a whole needs to decide how CRISPR-Cas9 should be used.

## CRISPR-Cas9

*CRISPR-Cas9 can be used to make precise changes to the genomes of living things. This could lead to new treatments for incurable diseases.*

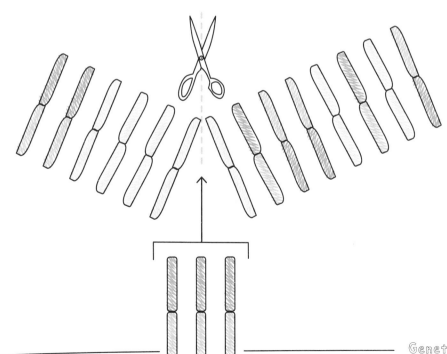

Adenine

Thymine

Cytosine

Guanine

# Who Am I?

There is no one like you. Everyone who has ever lived, is living, or will live on this planet, is unique. Even identical twins grow up to become different people. They have different personalities and habits, and often develop different diseases. We are all different, so what makes 'you' you?

Genetics plays an important role, but so does the **environment**. Deciding which of these two factors is more important is difficult. This is called the **nature-nurture** debate.

Scientists use **twin studies** to address this. Twin studies work by comparing pairs of identical and non-identical twins. Identical twins develop when a fertilised egg splits in two, so their DNA is the same. Non-identical twins develop when separate sperm fertilise separate eggs, so their DNA differs. Twins also tend to grow up in the same, shared environment. The idea is that if a characteristic, such as hair colour or intelligence, has a genetic component, identical twins will be more likely to share the characteristic than non-identical twins.

So far, there have been more than 2,700 twin studies, looking at more than 1,500 different characteristics. The results are clear. Whatever feature you look at — obesity, personality, mathematical ability — the test results of identical twins are always more similar than those of non-identical twins. Put simply, everything has a genetic component.

Our genes influence us, but our fate is not predestined. Twin studies also show that while genes are important, nothing is completely determined by them. Some features, such as height and eye colour, are more influenced by genetics or 'nature'. Other features, such as addiction and migraine, are more influenced by our upbringing

## Twin studies

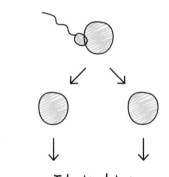
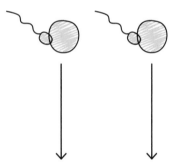

**Identical twins**
*share all of their genes*

**Non-identical twins**
*share about half their genes*

and the environment or nurture. This is called heritability. Heritability is a measure of how much differences in people's genes account for differences in their characteristics.

There is no gene for addiction or intelligence or any other complex feature. Instead, there are hundreds, or even thousands of different interacting regions in the genome that contribute to these complex characteristics.

When scientists decoded the genomes of 1,000 different people from all around the world, they found that the genomes of non-related people differ in about 3 million places. Our genetic codes are similar, but different. These differences or **variants** contribute to all of our idiosyncrasies. They are part of what makes us unique.

Now the nature-nurture debate has moved on. Scientists realise that nature and nurture do not act in isolation. They interact with one another. The environment can 'talk' to the body.

Environmental factors, such as diet, smoking and alcohol, can influence the way that genes work. This is called **epigenetics**. Epigenetic changes do not alter the DNA sequence. Instead, they alter gene expression. They do this by adding chemical labels, like Post-it Notes, to the DNA, signalling key genes to be switched **'on'** or **'off'**. Some even think that these epigenetic labels can be passed down the generations, so the lives that we lead may affect the gene activity of our children.

So, the environment shapes our epigenetic profile, which in turn influences **gene expression**, which in turn may shape our behaviour, lifestyle choices and health — our environment — and so it goes on. Our genes and the environment are intricately linked.

However, sometimes **chance** events can have a profound effect on the way we turn out. These events can be external, such as an accident, or internal, such as a throat infection. Random events can shape our biology and behaviour, which leads to epigenetic changes, which leads to changes in gene activity and so on. Human beings are the products of nature and of nurture, with a large dash of chance thrown in.

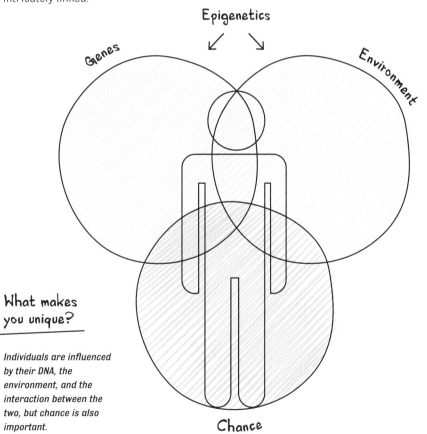

**What makes you unique?**

*Individuals are influenced by their DNA, the environment, and the interaction between the two, but chance is also important.*

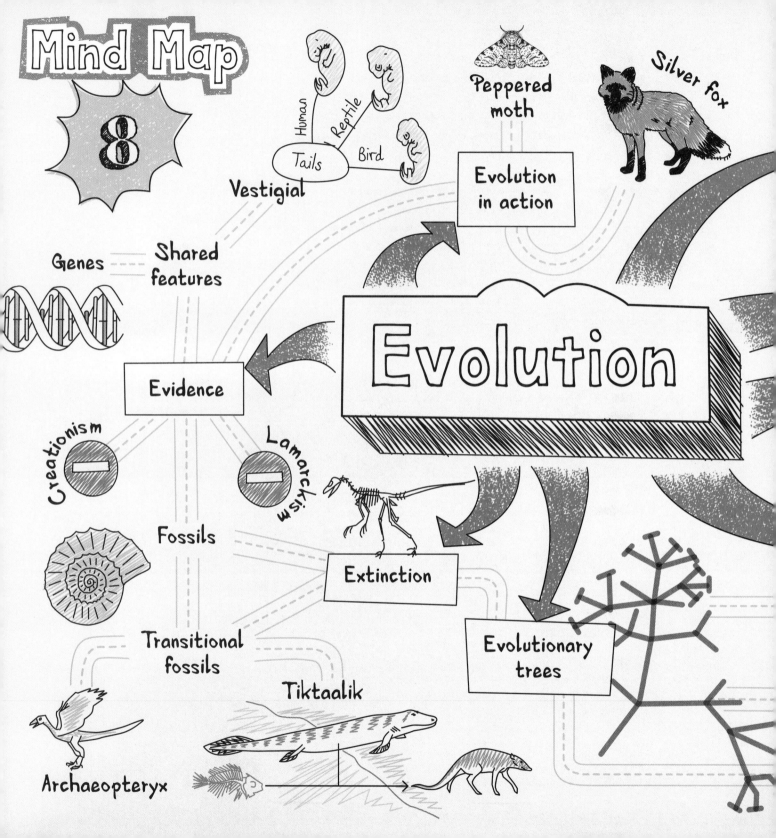

# Mind Map

**8**

Vestigial

Human
Tails
Reptile
Bird

Peppered moth

Silver fox

Evolution in action

Genes

Shared features

Evidence

# Evolution

Creationism

Lamarckism

Extinction

Evolutionary trees

Fossils

Transitional fossils

Archaeopteryx

Tiktaalik

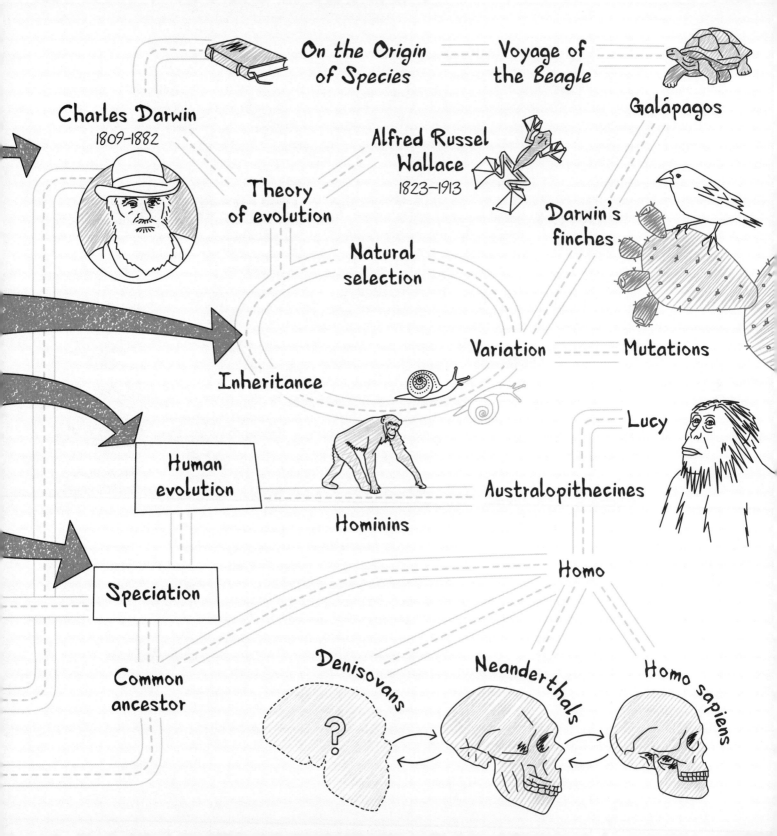

# Charles Darwin's Theory of Evolution

Evolution is the process by which living things change over time. The **theory of evolution** was proposed more than 150 years ago by the English naturalist **Charles Darwin**. Since then, it has become one of the most successful scientific theories ever. It helps to explain how the myriad different life forms that exist came to be, and how living things continue to change over time.

Darwin referred to evolution as 'descent with modification'. The idea is that new species evolve from pre-existing species, and that all species share a **common ancestor**.

He based his theory of evolution on a few simple ideas. Although individual members of the same species are similar, they are not exactly the same. There are subtle differences, called **variation**. In a single snail species, for example, some individuals may have spotty shells. Others may have plain shells.

### Variation

*Individuals show variation*

### Natural selection

*Spotty snails are better camouflaged and less likely to get eaten. They reproduce more*

### Inheritance

*Over time, spotty snails become more common*

The organisms that are best suited to their environment are the ones that are most likely to survive. Spotty shells may help to camouflage the snails, so they are less likely to get eaten by birds.

These organisms are then more likely to reproduce and pass their favourable characteristics on to their offspring. Meanwhile, individuals that are poorly adapted to the environment are less likely to survive and reproduce.

So, **inheritance** is a key feature of evolution. Snails with spotty shells are more likely to reproduce than snails with plain shells. Their offspring are likely to inherit this feature, so snails with spotty shells become more common over the generations.

Darwin called this idea evolution by **natural selection**. It causes change over time and helps organisms adapt to their environment. As the changes wrack up, new species emerge. This is called **speciation**.

In 1837, he sketched a structure in one of his notebooks and wrote next to it, 'I think'. Branching trees like this have since come to exemplify evolutionary theory. They show how one species can branch into different species, and how the process can repeat itself across time.

## The voyage of the Beagle

Darwin's ideas began to take form when he spent time aboard *HMS Beagle*. In 1831, he was invited to join the ship for a trip around the world. He was asked to be the captain's companion, and the ship's geologist and naturalist.

For the next five years, the *Beagle* surveyed the coast of South America, while Darwin explored the continent and its islands. He carefully observed the wildlife that he saw and made detailed drawings and written notes. He also collected thousands of specimens, which he packed up and sent home for further study.

When he arrived in the **Galápagos**, a group of volcanic islands located around 1,000 km (621 miles) west of Ecuador, he was amazed by the variety of species that he saw. He noticed that the species varied from island to island, and that although some species were superficially similar, they displayed certain key differences.

Some of the finches, for example, had narrow, pointed beaks that helped them to eat insects. Some had thick, curved bills that helped them to crack seeds, while others had long, sharp beaks that helped them to tear and eat cactus plants. They have since become known as **Darwin's finches**.

When he returned to England, Darwin thought about all the different species he had seen and all the different adaptations that they had. He then spent the next 20 years working on his ideas.

### Darwin's finches

*During his voyage on* HMS Beagle, *Darwin encountered the Galápagos finches. Their varied beaks provide evidence for his theory of evolution.*

In 1859, he published a book called ***On the Origin of Species***. It outlined his theory of evolution by natural selection. The book was an instant best seller. It sold out on its first day and has never been out of print since.

# Alternatives to Evolution

Darwin was not the first to try to explain how life on Earth emerged. In Victorian times, many people believed that God made the Earth and all living things in it. This is called **creationism**. According to the Bible, 'the creation' occurred a few thousand years ago, but now multiple lines of evidence tell us that this view is wrong.

By the start of the nineteenth century, scientists were coming up with alternative explanations. Jean-Baptiste Lamarck (1744–1829) was a French naturalist. He made an important contribution when he suggested that individual animals adapt and evolve to suit their environment.

According to his theory, the way an animal behaves during its lifetime has an effect on its body. So, if an animal uses one part of its body a lot, then that part of its body will get bigger and stronger. Any useful changes are then passed on to the next generation when the animal reproduces. Any features that aren't used, shrink and disappear. This is called **Lamarckism**.

For example, Lamarck suggested that giraffes acquired their long necks because successive generations of giraffes stretched up to reach the highest leaves. Stretching elongated the giraffe's neck, and this feature was then passed on to their descendants.

Although it has a certain childlike appeal, there are major flaws with Lamarck's theory. Firstly, it implies that as organisms become increasingly complex over time, simple organisms will disappear. This is clearly not the case.

Secondly, if Lamarck was correct, and offspring inherit features that their parents acquire during their lives, then the children of Olympic weightlifters should automatically inherit their muscly physique. They don't necessarily. They'll probably have to work out to build their big muscles, just like everyone else (see page 106).

It's now widely accepted that in almost all cases, Lamarck's version of evolution just doesn't happen, but his idea that life can change over time was a good one. It inspired Darwin to come up with his own theory of evolution.

## Lamarck's giraffes

*Lamarck thought that across generations the giraffe's neck becomes longer because they stretch to reach the tops of trees.*

# Wallace and the rise of Darwinism

Although Darwin's name is well known, he wasn't the only one to come up with the idea of evolution by natural selection. **Alfred Russel Wallace** was a British naturalist who lived round about the same time as Darwin, but worked separately. He went on trips to the Amazon Rainforest and the Far East, where he collected thousands of natural history specimens.

Like Darwin, Wallace wondered how the many different life forms that he witnessed had come to be. He realised that species exist in different forms, that better-adapted species will survive and breed, and that poorly adapted species will die out.

In 1858, he sent a scientific paper outlining his theory to Darwin. Darwin, who had yet to publish his own ideas, realised that the two men had arrived at identical theories. So, they decided to publish their work jointly. Two papers, authored by the separate men, were published by the Linnaean Society later that year. Both papers went largely unnoticed. Then Darwin published *On the Origin of Species* and his name became associated with the theory.

## Alfred Russel Wallace

*He may not be as well known as Darwin, but Alfred Russel Wallace came up with the theory of evolution independently after he studied the wildlife of the Amazon Rainforest and the Far East.*

Many people were excited by Darwin's theory, but others disliked it because it conflicted with the creationist point of view. Some people also thought that Darwin did not have enough evidence to back up his claims.

In the end, it took time for evolution to gain acceptance. By the time of his death, in 1882, Darwin was widely regarded as one of the world's most brilliant scientists.

# Evidence for Evolution

The **evidence** for evolution is all around us. One rich strand comes from the fossil record. **Fossils** are the remains of ancient organisms that are found preserved in rock, ice and other substances.

Fossils give us a peek into the lives of prehistoric organisms, and show us how living things change over very long periods of time. They are often found in rocks that build up in layers called strata. Because new layers of rock form on top of old, as successive layers of sediment are deposited, abseiling down a cliff face is like travelling back in time. Fossils in the upper layers are younger than fossils in the older layers.

The oldest rocks contain the simplest fossils, while the youngest rocks contain the remains of more complex organisms.

This supports the theory of evolution, which predicts that simple organisms gradually evolved into more complex ones.

Many weird and wonderful fossils have been found. These include everything from the remains of woolly mammoths and sabre-toothed cats, to the fossilised bones of dinosaurs such as *Tyrannosaurus rex* and *Triceratops*.

None of these species are alive today, so the fossil record teaches us that species come and species go. This also fits with the theory of evolution, which states that animals that are not well adapted to their environment are less likely to survive and reproduce than better-adapted animals. If a whole species is poorly adapted to a changing environment, then it is likely to become extinct. **Extinction** is part of evolution and a natural part of the story of life on Earth.

Although the fossil record is far from complete, scientists have used it to uncover a wealth of information, and new fossils are being discovered all the time.

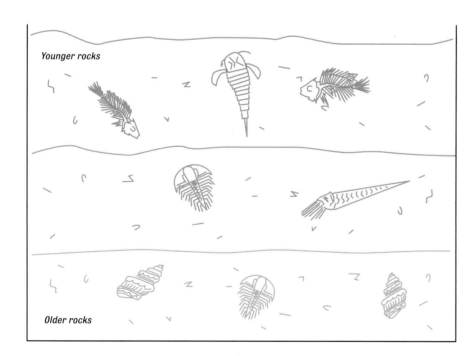

Younger rocks

Older rocks

Sometimes, stunning **transitional fossils** are found; these represent intermediate life forms and display a mixed bag of features. Some are shared with their ancestors, and some are shared with their descendants. They show **evolution in action**.

*Archaeopteryx* is one famous example. This raven-sized creature, which was discovered two years after Darwin published *On the Origin of Species*, had the wishbone, wings and feathers of a bird, but the teeth, tail and claws of a dinosaur. It is the transitional form between dinosaurs and modern birds, and it demonstrates that modern birds are descended from feathered non-avian dinosaurs.

*Tiktaalik* is another transitional fossil. It has scales on its back like a fish, but a flat head and neck like a land animal. Its fins are composed of bones that correspond to the bones found in all land animals. All life began in the oceans, then at some point, a creature crawled onto land. *Tiktaalik* may have been that creature. It looks like a fish that could walk, and researchers think it is the transitional form between bony fish and four-legged land creatures.

## Archaeopteryx

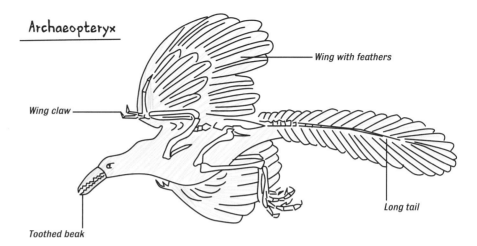

Wing with feathers

Wing claw

Long tail

Toothed beak

On rare occasions, the fossil record leaves an exquisite trace of a species' evolutionary history. The horse, for example, has a fairly complete fossil record. Multiple fossil remains reveal how a small, swamp-dwelling creature called *Hyracotherium* transitioned into the much larger modern horse over a period of around 50 million years.

The horse has changed considerably on its evolutionary journey, but the fossil record also shows that other animals change little. Crocodiles and sharks, for example, haven't changed much in millions of years. This is because they are well adapted to their environments, and because their environments have remained fairly constant too. This also fits with the theory of evolution.

## Evolution of the horse

*Detailed fossil remains reveal how the modern horse evolved from a much smaller, swamp-dwelling animal that lived 50 million years ago.*

*Modern horse*

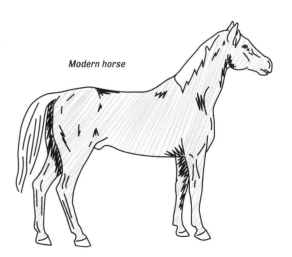

*Hyracotherium*

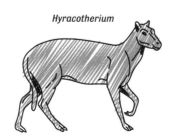

## Shared features

When Darwin proposed the theory of evolution by natural selection, he was unable to explain how variation occurs and how useful adaptations are inherited. No one knew about DNA, or genes or inheritance (see page 100).

Now, we understand the laws of inheritance and that variation is caused by **mutations**. Many of these random changes to the genetic code have no effect at all, but sometimes they do make a difference. Mutations can make individuals more or less adapted to their environment, and so influence whether an individual is likely to survive and reproduce. Mutations are the raw material that fuels evolution.

Geneticists have compared the genomes of many different species and found that organisms share many of their **genes**. Thousands of our own genes, for example, are also found in other organisms, such as bacteria and plants. This is because we all inherited them from a very distant common ancestor.

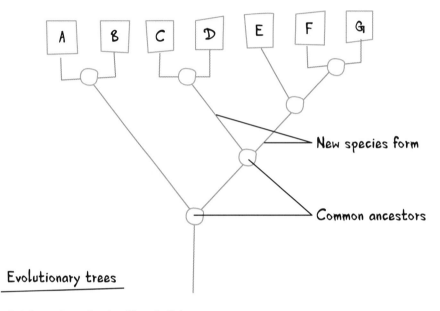

### Evolutionary trees

*Evolutionary trees show how life on Earth is related and how all life forms are descended from the same common ancestor.*

Shared features, such as these, are further evidence of evolution. On closer inspection, it becomes clear that some groups of species have more genes in common than others. For example, we know that humans share more genes with apes, such as gorillas and chimpanzees, than we do with other species – around 96%. This suggests that apes are our closest relatives; a finding that is backed up by detailed analyses of the fossil record.

Now researchers draw precise **evolutionary trees** to represent how organisms are related. Branches are places where new species have formed, and converging lines show the common ancestors that gave rise to various organisms. These are based on multiple lines of evidence, including the fossil record and DNA evidence. The primate evolutionary tree suggests that humans and chimpanzees shared a common ancestor around 5 to 8 million years ago.

The theory of evolution predicts that closely related species will share similar features, and this is exactly what is found. What's a little odd, however, is that sometimes distantly related species share unexpected features.

For example, the forelimbs of all mammals, including humans, dogs, whales and bats, have the same arrangement of bones even though their forelimbs do very different things. Humans use theirs for lifting, dogs use theirs for walking, whales use theirs for swimming, and bats use theirs for flying.

### Forelimbs in mammals

*The forelimbs of different mammals all have a similar bone arrangement, which they inherited from a common ancestor.*

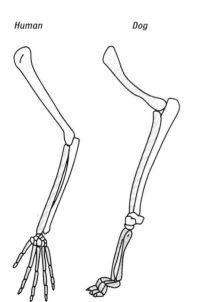

| Human | Dog | Whale | Bat |

These unusual characteristics only make sense in the context of evolution. The different species have the same bone arrangement because it was present in their common ancestor millions of years ago.

It's a similar story for **tails**. All vertebrate embryos, from humans to birds to fish, have a tail at some point during their development. This is because the common ancestor of vertebrates also had a tail.

**Vestigial** features such as tails have little or no purpose. They are just the remains of a feature that used to serve a purpose in the evolutionary past.

### Vestigial feature

*The tissue fold in the corner of your eye is a vestigial feature inherited from distant ancestors.*

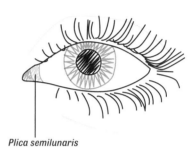

*Plica semilunaris*

The small fold of tissue that lines the inside corner of the human eye is another vestigial feature. Some snakes also contain the remains of a pelvis and leg bones, because they evolved from a four-legged ancestor that used to walk on land.

Although complex features are influenced by complex combinations of genes, sometimes it's possible to pinpoint the exact mutation that caused a significant evolutionary change. In 2016, for example, scientists identified a mutation in a stretch of snake DNA called ZRS. This one small change was enough to rid the animals of their limbs and confine them to a future of slithering on their bellies.

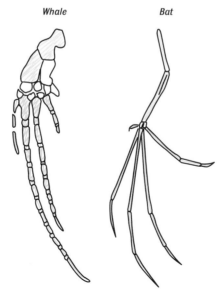

# Evolution in action

Although evolution tends to happen slowly, across thousands or millions of years, sometimes it's possible to see evolution in action. The **peppered moth** is a famous example. Before the nineteenth century, the UK's peppered moths were cream coloured with little black spots. Then, in 1848, a moth collector from the city of Manchester caught a sooty-coloured individual. In the years that followed, this 'melanic' version of the moth became increasingly common, until by the turn of the century, it had all but replaced the usual, lighter form in industrial areas.

The cause was the Industrial Revolution. Tall chimneys belched out thick clouds of smoke, and the soot settled on the tree trunks where the moths liked to rest during the day. The cream-coloured moths were at a disadvantage. The birds spotted them easily and gobbled them up.

The melanic moths, in contrast, were perfectly camouflaged against the dirty background. They were less likely to get eaten, and more likely to survive and reproduce. Over time, because melanism is heritable, the number of sooty-coloured moths increased in urban areas, and the number of lighter coloured moths declined.

Then when the Clean Air Act was introduced to spruce up the skies in the mid-twentieth century, things changed again. The melanic form declined in urban areas, and the lighter form became more common.

The story of the peppered moth clearly shows that species change over time in response to their environment. This is a classic example of evolution by natural selection, and in 2016, scientists deciphered the nature and timing of the genetic change that was responsible.

A fortuitously timed mutation, which appeared in 1819 toward the start of the British Industrial Revolution, altered a gene that controls the moth's colouring. It changed the fate of the peppered moth, and was enough to turn this unassuming insect into an evolutionary icon.

## Peppered moth

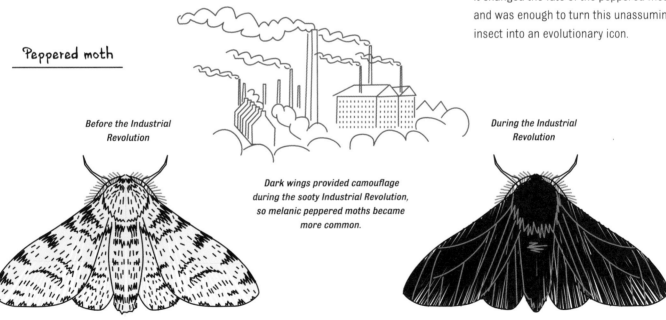

*Before the Industrial Revolution*

*Dark wings provided camouflage during the sooty Industrial Revolution, so melanic peppered moths became more common.*

*During the Industrial Revolution*

From silver fox to tame fox

The original
silver fox

The tame fox
after artificial
selection

## ARTIFICIAL SELECTION

An unusual experiment in the wilds of Siberia also demonstrated how species could change over time, and involved the **silver fox** as its subject. In 1959, Russian geneticist Dmitri Belyaev took 130 silver foxes from local fur farms, and set up a breeding experiment.

The foxes were wild animals. Most of them snarled and lunged at him, but a small subset was slightly less aggressive. So Belyaev took these animals and bred them together. Then when their offspring grew up, he took the least aggressive adults and let them have pups. The process was repeated. Every generation, Belyaev allowed only the tamest individuals to breed.

Change came thick and fast. Within a few generations, the foxes started whining, whimpering and wagging their tails. The proportion of friendly pups grew in each generation, until by generation 45, virtually all of the foxes were acting like friendly dogs.

They looked different, too. Some developed spotty markings, just like those sported by some domestic animals. Their legs became shorter, their snouts became smaller, and their skulls became wider. The foxes were evolving into something less fox-like and more dog-like.

There was variation in the original group of founding animals, then Belyaev deliberately let only key animals reproduce. Instead of natural selection, where species respond to natural changes in the environment, this is artificial selection, where species respond to artificial changes in the environment that are caused by humans.

Just like the peppered moth, the silver fox study shows that evolution doesn't have to take thousands of years. Even a slow-breeding animal like a fox can change substantially in just the space of one human lifetime. It also shows how just selecting for one feature – tameness – can cause behavioural and physical changes, and hints that our ancestors may have followed a similar strategy when they began to domesticate wild animals many thousands of years ago.

# Human Evolution

Although humanity's reach is global (see page 139), we are just one tiny twig on the evolutionary tree of life. Like all living things, we can trace our ancestry back to the first, single-celled creatures that lived on Earth over 3 billion years ago, but our story doesn't really get going until much more recently.

Around 7 million years ago, humans and chimps shared a common ancestor. After that, some of the descendants evolved into modern chimps, but others went on to become human. These were our earliest human relatives. We call them, and all of their varied descendants, **hominins**. These early hominins lived in Africa, and showed a mix of ape-like and human-like features.

The ***australopithecines*** were the first to emerge around 4.4 million years ago. The most famous is **Lucy**, a female australopithecine fossil that was found near what is now Hadar, in Ethiopia.

Although she has a small chimp-sized skull, her body is less chimp and more human. Unlike her ancestors, Lucy stood tall and walked on two legs. This is a milestone in the story of human evolution. Australopithecines were the first hominins to leave the forest and live on the savannah.

From 2.4 to 1.5 million years ago, our genus, ***Homo*** appeared. We know of at least nine different types of Homo. They include *Homo habilis*, *Homo erectus* and *Homo neanderthalensis* (**Neanderthals**). Many of these overlapped in time (see page 12).

Unlike earlier hominins, who exhibited a mix of human and ape-like features, these *Homo* species were far more human-like. They had bigger brains than their ancestors. They were the first to develop sophisticated stone tools, the first to control fire, and the first to migrate out of Africa and explore the rest of the world.

Modern humans – ***Homo sapiens*** – are the most recent addition to the *homo* genus. It's estimated that we evolved around 300,000 years ago.

As new fossils are unearthed, and their DNA is probed, the complexity of this story is becoming increasingly apparent. Genetic studies have shown, for example, that modern humans and Neanderthals interbred in Europe tens of thousands of years ago.

If you are of European or Asian descent, then 1–4% of your genome is made up of Neanderthal DNA. This includes genes involved in the immune system and in skin and hair colouring. Natural selection has favoured their persistence because they have been useful to us.

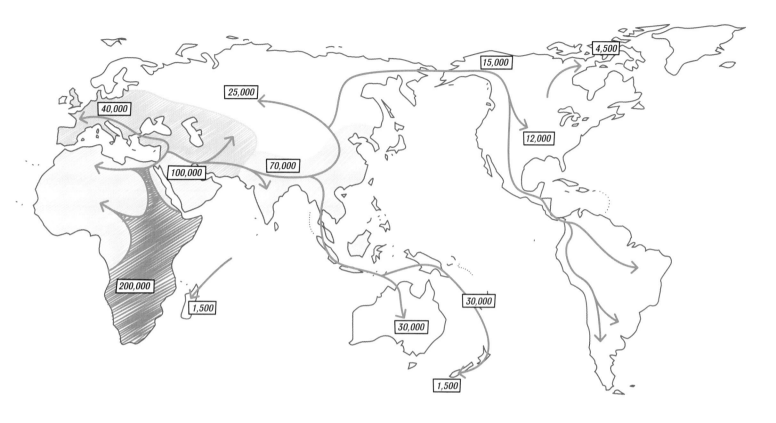

Genetic analyses also show that Neanderthals interbred with the **Denisovans**; an extinct species of hominins that lived in parts of East Asia. Species overlap, move around and interbreed. As we find out more about our ancestors, we're finding that the human evolutionary tree is less of a tree and more of a thicket!

New species evolve, and older species become extinct. Out of 20 or so known types of hominin, we are the only species left, and for as long as our species remains, it continues to evolve.

Thousands of years ago, for example, we lost the ability to synthesise vitamin C, and gained the ability to digest the lactose sugar in milk as adults.

New mutations are cropping up all the time. They provide the variation that powers natural selection, and help humans – and all other life – to adapt to a changing world. Evolution is an ongoing process that never stands still.

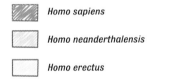

Homo sapiens

Homo neanderthalensis

Homo erectus

## How modern humans dispersed

*Around 200,000 years ago, early humans migrated out of Africa and slowly spread across the globe.*

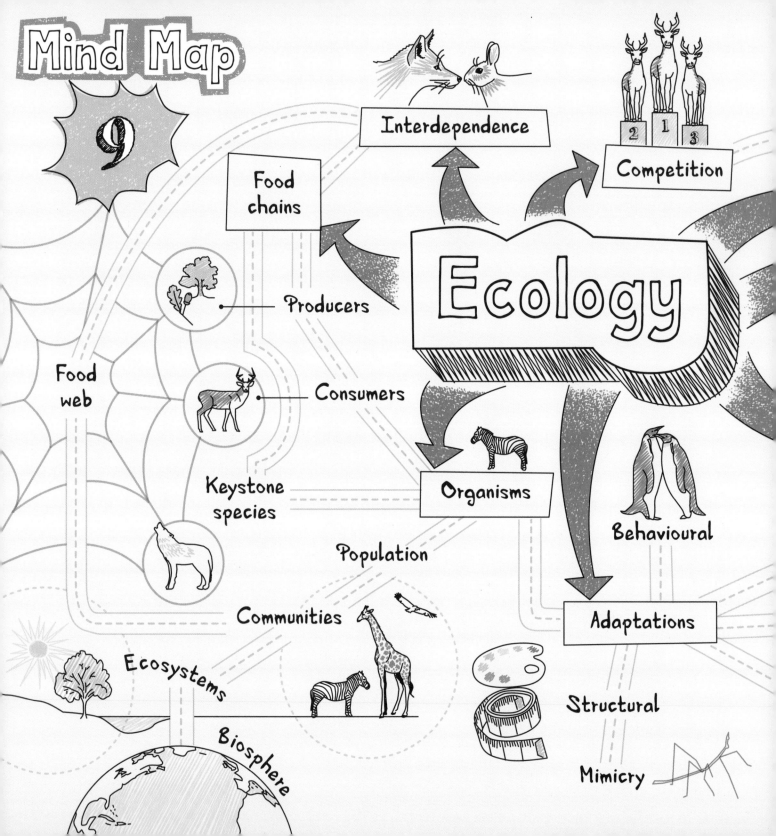

# Mind Map

## 9

**Interdependence**

**Competition**

**Food chains**

**Ecology**

Producers

Consumers

Food web

Keystone species

Organisms

Behavioural

Population

Adaptations

Communities

Ecosystems

Structural

Biosphere

Mimicry

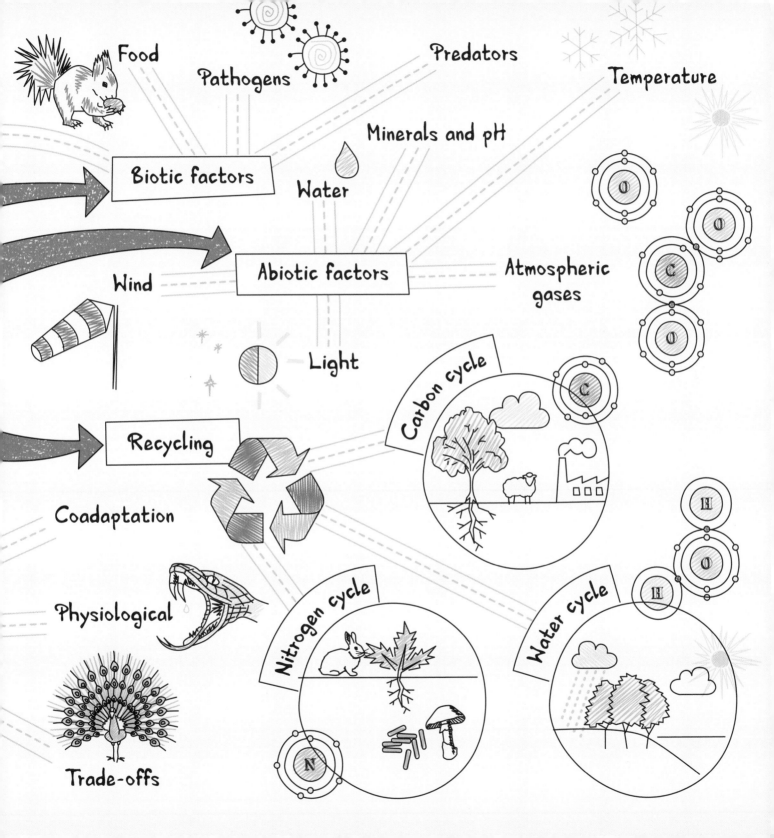

Food

Pathogens

Predators

Temperature

Minerals and pH

Water

Biotic factors

Abiotic factors

Atmospheric gases

Wind

Light

Carbon cycle

Recycling

Coadaptation

Physiological

Nitrogen cycle

Water cycle

Trade-offs

# The Study of Ecosystems

Living things do not live in isolation. They interact with each other and with their physical environment. **Ecology** is the study of these relationships.

The field of ecology has become increasingly important as humanity's impact on the planet has grown. It is important to understand how living things interact with their environment, so that we can work out how to minimise the negative impacts that our actions are having on the planet.

Ecology is studied at many different levels. The simplest level is that of the individual organism. Here, ecologists study useful features that have arisen due to natural selection (see page 110). These are characteristics, such as the patterning on a butterfly's wing or a cheetah's ability to run quickly, which help the organism to live in a particular environment.

At the **population** level, ecologists study groups of **organisms** that belong to the same species and that live together. Scientists are interested to find out how the size, density and structure of various populations changes over time.

Community-level ecology studies how populations of different species interact and live together. **Communities** frequently include plants, animals and other smaller organisms, such as bacteria. The members of these communities depend on one another and are adapted to the setting that they live in.

At the next level, ecologists study **ecosystems**. Ecosystems are made up of communities of living things and their physical environment.

The Earth is made up of many different ecosystems, so the final level of study is at the planetary level. The **biosphere** is all of our planet's ecosystems and landscapes added together. Global ecology examines how planetary-scale influences, such as climate change, influence the functioning and distribution of organisms across the biosphere.

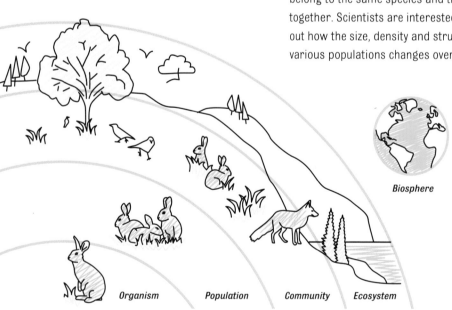

Biosphere

Organism    Population    Community    Ecosystem

### The scope of ecology

*Ecology operates at many different levels, from the level of a single organism, right up to the planetary scale.*

# Interdependence

All of the organisms inside an ecosystem depend on each other. This is a key feature of ecology.

Animals depend on plants, as they make oxygen for animals to breathe. They provide shelter for animals to hide in, raw materials for building, and food for energy. Plants depend on animals for pollination, nutrients and seed dispersal. When they die and decompose, animals become fertiliser that helps plants to grow.

Living things are part of **food chains**. Food chains are sequences of interconnected organisms, where one organism feeds on another organism, which feeds on another organism, and so on. For example, foxes eat rabbits, which eat grass.

If one of the links in the food chain is disrupted, then the status quo changes. For example, if all of the foxes were killed, the number of rabbits would increase and the amount of grass would decrease.

Often, small changes to an ecosystem can have big repercussions. These consequences are not always easy to predict. We call this interrelatedness **interdependence**.

If the environmental factors do not change much, then an ecosystem may find it remains relatively stable. The number of species stays roughly the same, as do the population sizes of the various species. These are called stable communities. They occur when populations live in balance with the environment. Stable communities are very important.

Mature coral reefs, oak woodland and tropical forests are examples of stable communities. Although the environment alters as the seasons change, these patterns are regular and more or less predictable. Communities like these harbour a wide range of other species. When a large stable community, such as a coral reef or a tropical forest, is lost, it is not easily replaced.

## Food chains

*When one element in a food chain changes, it has knock-on effects for all the other component parts.*

# Living Things and Their Environment

Living things are influenced by biotic and abiotic factors. **Biotic factors** are related to living things. **Abiotic factors** are non-living things. Biotic and abiotic factors affect individual organisms and their communities.

## Biotic factors

There are four main types of biotic factors: food availability, pathogens, predators and competition.

1. **Food**: Living things need energy, which they obtain from food. The amount of available food influences the number of animals that an ecosystem can support. So, tropical rainforests, which contain many, varied food sources, support far more species than deserts, which have fewer food resources.

2. **Pathogens**: When disease-causing organisms invade new territories, the animals that live there often become sick because they have no resistance to the disease. New pathogens can wipe out entire populations. For example, the chytrid fungus, which is spreading around the globe, is decimating the world's amphibian populations and may have driven some species to extinction.

3. **Predators**: When new predators arrive in an environment, the effects can be devastating. After European settlers introduced cats to New Zealand in the nineteenth century, the cats went feral and started to eat the local birds. Since then, they have driven at least six native bird species to extinction, and caused the decline of many resident populations.

4. **Competition**: Species sometimes compete with each other, and sometimes one species can outcompete another. Japanese knotweed is a classic example. This tall plant is native to parts of East Asia, but in many other parts of the world, it is outcompeting the native plants. It's now labelled as one of the world's worst invasive species.

## Living influences

*Living things are influenced by biotic elements that relate to living things.*

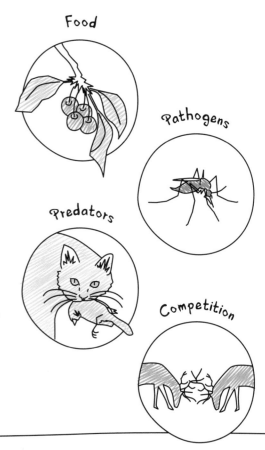

Food

Pathogens

Predators

Competition

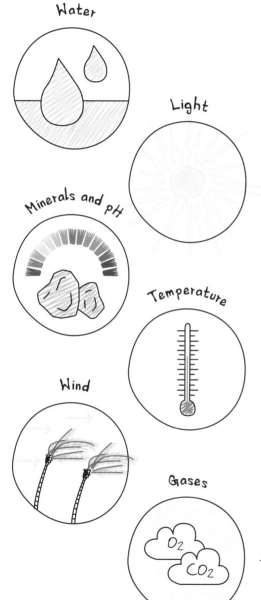

Water

Light

Minerals and pH

Temperature

Wind

Gases

## Abiotic factors

Abiotic factors are equally important to an organism or community's survival. There are six important non-living abiotic factors: atmospheric gases, light, temperature, water, mineral content and pH, and wind.

1. **Atmospheric gases**: Plants need carbon dioxide for photosynthesis, and animals need oxygen for aerobic respiration (see pages 60–63). Aquatic creatures are particularly sensitive to the oxygen concentration in the water. Some invertebrates can survive with very low oxygen levels, but fish tend to need much higher levels of the dissolved gas.

2. **Light**: The intensity of light has a big effect on plants, because they need light for photosynthesis. Some plants like bright light, while others are adapted to grow in the shade. They may have bigger leaves or more chlorophyll. Animals respond to light, too. The daily habits and breeding cycles of many animals are related to day length and light intensity.

### Non-living influences

*Non-living influences, such as temperature and water availability, also affect living things.*

3. **Temperature**: Animals and plants have evolved to grow well at their optimum temperatures. Cold temperatures are always a problem. In the Arctic, for example, plants are often stunted and small. This limits the number of herbivores that can live there, as well as the number of predators who feed on them.

4. **Water**: Plants and animals need water to survive. Without water, there is little or no life. Some plants, such as cacti, are adapted to low water levels, while others, such as pitcher plants, grow well in moist, boggy conditions.

5. **Minerals and pH**: Some plants, such as rhododendrons, like acidic soil, while others, such as lavender, like soil that is more alkali. Organisms also need certain minerals to grow. Plants need nitrates to help them make amino acids and magnesium to help produce chlorophyll.

6. **Wind**: The intensity and direction of wind affects many organisms. Most organisms prefer sheltered conditions. Those that live in windy places, such as sand dunes, have evolved useful adaptations. Some dune-dwelling plants, for example, grow as mats. This helps to anchor them to the ground and each other.

# Adaptations

L ife on Earth can be found just about everywhere, from the polar ice caps to the tropical rain forests, and from deep-sea vents to the heights of the stratosphere. Living things have special features that help them to survive in the conditions where they live. These features are called adaptations.

Adaptations emerge via the process of evolution (see page 110). The adaptations that persist are the ones that help organisms to survive and reproduce.

Adaptations can be structural, behavioural or physiological. **Structural** adaptations are physical features, such as the size, shape or colouring of an organism. **Behavioural** adaptations are inherited behaviours, such as mating or searching for food. **Physiological** adaptations enable the organism to do certain things, such as making venom or spinning a web. General features such as growth, development and regulating temperature are also physiological adaptations.

### Asian sand cat

*This desert-dwelling feline has many adaptations to help it survive in harsh, hot conditions.*

The Asian sand cat lives in the desert belts that stretch from North Africa, through the Middle East and up into central Asia. It is the only cat to live exclusively in desert habitats, and it has evolved a number of specialised adaptations that help it to live there.

It avoids the worst of the heat by sleeping during the day in cool, underground burrows. This is a behavioural adaptation. It employs camouflage to blend in with its background; its sandy coat helps it to sneak up on its prey. It has stocky limbs to help it dig, large ears to help it hear, and a powerful bite that crushes its prey. Uniquely, the undersides of its paws are completely covered in fur. This helps it to move around on the sand. These are all structural adaptations.

It doesn't really need to drink and gets most of its moisture from its food sources. This is a common physiological adaptation in desert-dwelling animals. The kidneys produce a small amount of highly concentrated urine, enabling the animal to retain as much water as possible.

Animals that live in cold places face different challenges. They need to stay warm. Insulation is important, so many animals that live in the cold have thick layers of blubber and/or dense fur coats. The surface area to volume ratio is also important. If an animal has a small surface area to volume ratio, it helps to reduce heat loss. This is why so many Arctic mammals, such as walruses, whales and polar bears are relatively big.

Some species go a step further. Some Antarctic fish have evolved the ability to make antifreeze proteins in their blood. This literally stops them from freezing when the temperature drops. The North American wood frog takes a different approach. It lets itself freeze, but has evolved a way to protect its cells from the damage that freezing and thawing causes.

**Mimicry** is another adaptation. This is where one species evolves to look, sound or smell like another. Birds use sight to tell the difference between tasty and noxious insects, so some palatable insect species have evolved to look like noxious ones. For example, some hoverflies have evolved to look like wasps or bees. The Chinese character moth avoids being eaten because it has evolved to look like a bird dropping.

## Hoverfly

*Their wasp-like appearance helps hoverflies avoid predation.*

## Coadaptation

*The tree shrew and the pitcher plant have evolved complementary adaptations.*

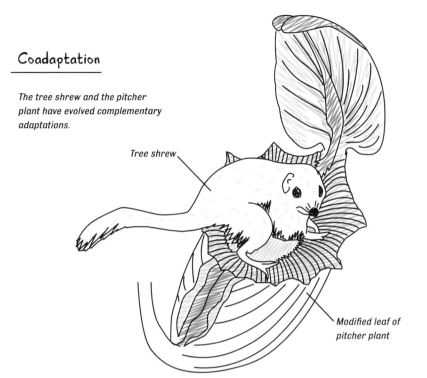

Tree shrew

Modified leaf of pitcher plant

Sometimes species that evolve alongside each other develop complementary adaptations. This is called **coadaptation**. In the mountainous cloud forests of Borneo, the tree shrew and the pitcher plant have coadaptations. The plant's modified leaves form deep pitchers with little lids. They act as toilets for the tree shrews that hop on them, lick nectar from the underside of these 'lids', and then defecate into the elongated receptacles. The shrew receives a much-needed nutrient boost, and the plant uses the faeces as a source of nitrogen. This is an example of mutualism because both species benefit from this unusual interaction.

Adaptations often involve **trade-offs**. The giraffe's long neck, for example, is a formidable weapon and helps the animal to graze from tall trees, but it's also heavy and takes energy to maintain. The peacock's ornamental train helps to attract females, but it's conspicuous and unwieldy. The elongated bodies of stream-dwelling salamanders help them to manoeuvre in the water, but increase their risk of drying out. Adaptations come with pros and cons. Evolution doesn't come up with perfect solutions. Instead, compromise is everywhere.

# Competition

### Pride of lions

*Lions live in prides and defend their territory. Their roars travel long distances, warning others to steer clear.*

Species come into competition when their roles are similar and they both depend on the same resources. Competition can occur between members of the same species, or between members of different species. If one species competes against another and is unsuccessful, then that species may become extinct. This is a natural part of evolution (see page 114).

## Competition in animals

Animals compete for food, territory and mates. Food provides animals with energy and the raw materials to carry out cellular processes (see page 54). Animals die without adequate food, so competition for food can be fierce.

Herbivores eat plants. Some herbivores eat only one kind of plant, while others eat different kinds. The caterpillar of the painted lady butterfly feeds on a wide range of food plants, including thistles and nettles, while the caterpillar of the chalkhill blue butterfly feeds only on horseshoe vetch.

Being a fussy eater is a risky strategy, because the population may die out if the food disappears. Having a broad diet is a good idea, but it means that other animals with similar tastes may compete for it. Nettle is a common food plant for many different butterfly species.

Carnivores are meat eaters. They often compete for prey. In the United States, coyotes, foxes and bobcats all compete for the same prey: small mammals. If they can, coyotes will kill and eat their non-coyote competitors.

Many animals go to great lengths to establish and defend their territory. Lions live in prides with a few adult males, related females and cubs. Each pride has its own territory. The lion's roar is a territorial display that can be heard up to 8 km (5 miles) away. It warns others to stay away.

Territories are important because they provide vital resources, such as food and a place to raise a family. Many animals, including lions, use their urine or faeces to mark the boundaries of their territories.

Competition for mates is also fierce. Some animals literally compete head on. Male red deer lock antlers, and the dominant, successful male then gains access to the females. Others take a less physical approach. Kakapos are extremely rare, ground-dwelling parrots that live in New Zealand. As the mating season approaches, males fashion bowls in the ground with tracks leading up to them. Then they stand in their bowls and make repetitive booming noises. Females listen to the songs and then mate with the male with the best boom.

### Kakapo

*Male kakapos compete for females by booming from their bowl-shaped stages.*

## Competition in plants

Just like animals, plants sometimes have to compete with members of their own species and members of other species. Plants compete for food, light, water and space to grow.

Big, tall plants use many resources, such as minerals and water. They also reduce the amount of light that reaches plants growing nearby. Neighbouring plants need adaptations to help them with these issues.

One way to deal with this competition is to flower at different times. In British woodlands, small plants such as snowdrops and bluebells flower early in the year, before the tall trees around them have established their canopies. This gives them access to light, while the decomposing leaves on the forest floor provide a ready source of minerals.

Another way to compete is to grow taller or have big leaves with a large surface area. Both strategies help the plant to make the most of the available light and help to fuel growth via photosynthesis. Some plants avoid competition for water by having long, deep roots that can access water deep underground, whilst others have a broad network of shallow roots that tap into water supplies nearer the surface.

Sometimes, a tree may find itself in competition with its own offspring if its seeds fall nearby. So, plants have evolved strategies to spread their seeds. The seeds of the maple tree are known as helicopters or whirlybirds. They are shaped to spin as they fall, and be carried by the wind.

Carnivorous plants often grow in nutrient-poor places, like acidic bogs, where they compete with other plants for these scarce resources. Carnivory is an extreme adaptation that helps the plants to survive. The Venus flytrap famously snares insects between the jaws of its modified leaves. Trigger hairs inside the trap respond to touch, so when a flying insect lands, the jaws of the trap snap shut. The whole process takes less than a second.

# Food Webs and Keystone Species

The world is full of organisms that eat other organisms; they are called **consumers**. Organisms that make their own food, such as plants, are called **producers**. Food chains always start with a producer and end with a consumer. They can be represented as a pyramid of numbers. This is a type of bar chart (see below). Each level in the bar chart represents a population from the food chain. The bars are drawn to scale, so a wider bar means that more organisms are present.

## Food web

Food web diagram: Grass → Rabbit, Insect, Slug. Insect → Frog, Shrew, Songbird. Rabbit → Fox. Frog → Hawk. Shrew → Hawk. Songbird → Hawk. Hawk → Fox.

Often these diagrams look like pyramids, but sometimes the overall shape is less obvious. This can happen if the producer at the bottom is a large plant, or if one of the animals in the pyramid is very small.

Food chains are useful because they help us to see some of the relationships that exist between species, but in reality, they're far more complicated than just 'fox eats rabbits eats grass'. Food chains interact with one another, and if you join them all up together, you get a **food web** (see above).

## Food chain

Pyramid of numbers from top to bottom: Hawk, Shrew, Grasshopper, Grass.

Food webs show how sometimes, prey can have multiple predators, and predators can have multiple prey. This can lead to interesting situations. If one type of prey dies, for example, then the predators may survive by eating another. This can have repercussions for the rest of the food web.

**Keystone species** are key parts of these food webs. A keystone species is any species of plant or animal that plays a unique and crucial role in the functioning of an ecosystem. They are named after the keystone in a bridge. If the keystone is removed, the bridge crumbles. If a keystone species is removed from an ecosystem, then the ecosystem either crumbles or becomes radically different. The wolf is a keystone because it influences lots of other species, such as coyote, antelope, elk, beavers and birds.

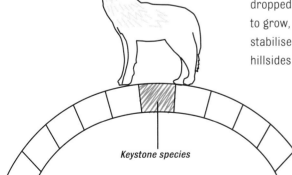

*Keystone species*

## The wolves of Yellowstone National Park

Wolves are a good example of a keystone species. In the early twentieth century, all of the wolves in Yellowstone National Park in the United States were exterminated because farmers feared for their livestock. In the years that followed, the ecosystem changed as the remaining members of the wolf's food web adjusted to the change.

Without wolves to compete with them, the coyote population increased. The coyotes fed on the pronghorn antelope population, so their numbers decreased. With no wolves to eat them, the elk population soared and species that the elk ate – deciduous trees such as aspen and willow – became severely overgrazed.

This meant there was less wood for beavers to build their dams and fewer places for birds to nest, so in turn, their numbers declined. With fewer beavers and fewer beaver dams, the water table dropped. This made it harder for trees to grow, and with fewer tree roots to stabilise them, the riverbanks and hillsides began to erode.

Then, in 1995, the authorities decided to reintroduce wolves back into Yellowstone. Fourteen animals were released, and the ecosystem changed again.

The elk learned to stay away from the valleys and ravines where the wolves could trap them, so these areas began to regenerate. Saplings sprouted and trees grew. Their roots helped to bolster the riverbanks, so they eroded less often and the rivers became more fixed in their course.

Birds and beavers came back. The beavers built dams and created ponds that provided habitats for fish, amphibians and reptiles. The wolves killed coyotes, so the numbers of small mammals increased. This led to an increase in hawks, weasels and foxes.

Biodiversity rocketed, all because one single species was allowed to return. Now conservationists realise the value that keystone species have to remodel not just landscapes, but the species that live in them.

# Recycling

The planet has been recycling things for billions of years. All life on Earth is made up of the same basic building blocks; for example, carbon, nitrogen and water. Living things remove these materials from the environment and use them for vital processes such as growth, but they are subsequently returned to the environment to provide the building blocks for future life. The nitrogen cycle, the carbon cycle and the water cycle are three very important global recycling schemes.

*Nitrogen is constantly recycled by the actions of bacteria, fungi, plants, and animals.*

Nitrogen in the atmosphere

Nitrate

Nitrite

Decomposers (bacteria and fungi)

Ammonia

## The nitrogen cycle

Living things need nitrogen to help build amino acids and nucleic acids (see page 27). Nitrogen makes up around 80% of Earth's atmosphere. Living things cannot use nitrogen in this form, so it has to be converted into biologically useful versions.

Bacteria and other single-celled creatures do this. It's called nitrogen fixation. Nitrogen-fixing microbes convert nitrogen gas into ammonia, which can then be taken up by plants. Some of the ammonia in the soil is modified further. Bacteria in the soil convert ammonia into nitrite and then nitrate. This is called nitrification. Plants can also take up these simple nitrogen-containing molecules.

Plants use these nitrogen-containing substances to make proteins and fuel their growth. When animals eat plants, they acquire this nitrogen. Animals also use nitrogen to help make proteins and help them grow, but it doesn't stay in their bodies forever. Some is expelled in the form of urine and faeces. Then when the animal dies, the organic nitrogen in its body is converted back to nitrogen gas. This involves more organisms and often takes several steps.

The organisms that break down dead animals and plants are called decomposers. The process starts when detritus feeders, such as maggots and worms, break down these dead organisms and release waste materials. Smaller organisms, such as bacteria and fungi, then digest everything. They use some of the nutrients to grow and reproduce, and produce carbon dioxide, water and mineral ions as waste products.

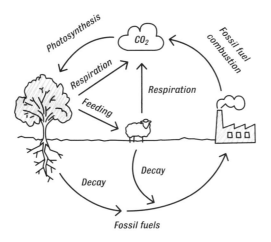

*Carbon is constantly recycled between living things and the environment.*

# The carbon cycle

Carbon is another vital element for life on Earth. All life on Earth is carbon-based, but there is also a lot of carbon in the environment.

Some of it is combined with other elements in carbonate rocks, and some of it is locked up in fossil fuels. Fossil fuels are formed from the remains of organisms that died millions of years ago.

There is also carbon dioxide gas ($CO_2$) in the air. Carbon dioxide is a product of respiration, but it also enters the atmosphere as a by-product when humans burn fossil fuels. Living things also return organic carbon to the environment when they produce waste products and when they die.

Balancing this out are plants and green algae that remove carbon dioxide from the atmosphere via photosynthesis. This carbon is then passed on to animals via food chains and food webs.

# The water cycle

Water is another vital compound that life depends on. The Sun turns liquid water into vapour. This is called evaporation. The water vapour rises into the air, and as it cools and condenses, it forms clouds. The wind blows the clouds over the land, and the water is returned to the earth when it falls as rain, snow, hail or sleet.

Some water runs into streams and rivers and is returned to the sea. Some seeps into the ground. Ground water is stored in underground rocks called aquifers. People all over the world rely on these aquifers for their fresh water. Some of the water remains in shallow layers of soil, where it is accessible to plants.

Plants draw up water from the ground. This is called transpiration. By allowing water to evaporate from their leaves, plants can continually draw up water from the ground. As animals and plants respire, they return water vapour directly to the atmosphere. Animals also return water to the environment in urine, faeces and sweat (in mammals).

*Water is always on the move. It evaporates into the air, falls as rain and returns to the sea. Then the process starts over.*

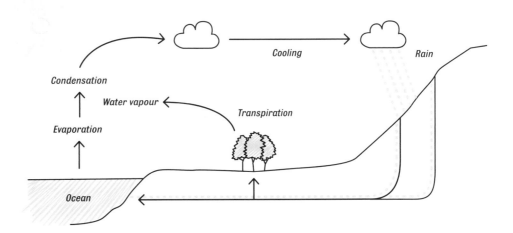

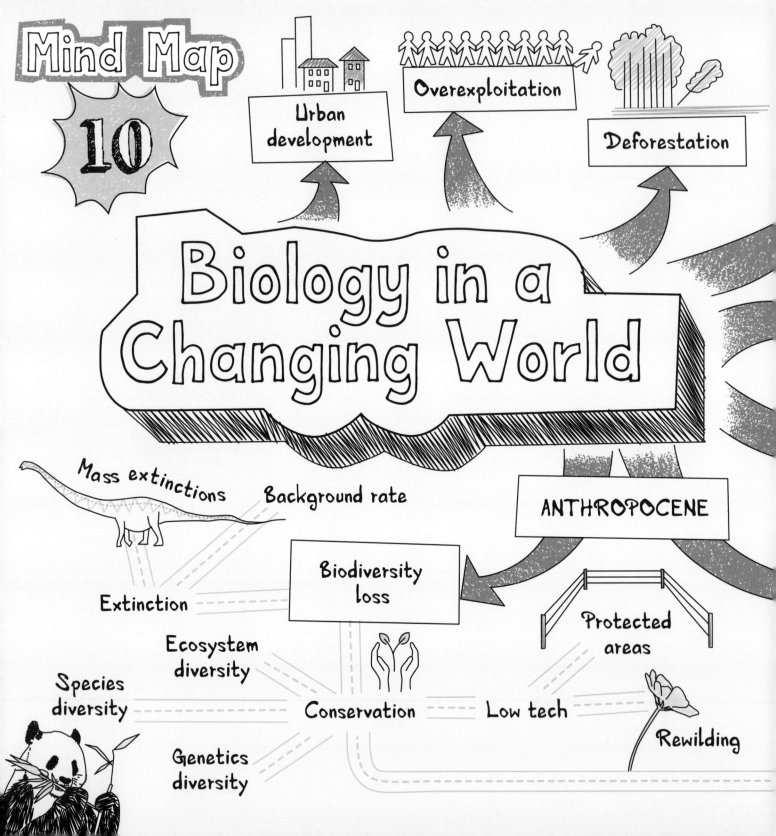

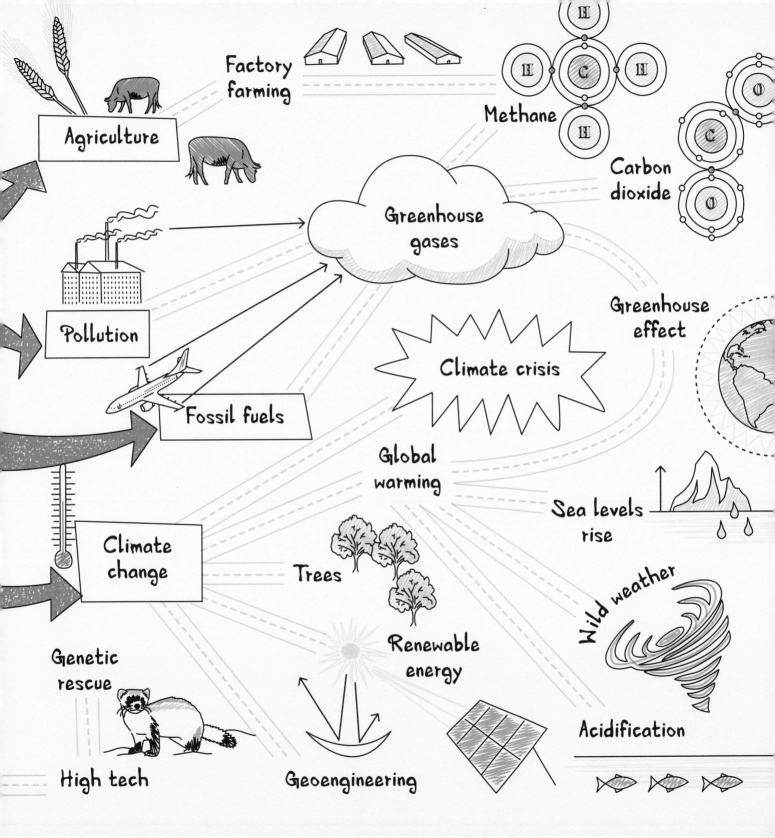

# Life in the Anthropocene

| | | |
|---|---|---|
| Anthropocene | | |
| Holocene | 11,700 years ago | Cenozoic |
| Pleistocene | 1.8 million | |
| Pliocene | 5.3 million | |
| Miocene | 23 million | |
| Oligocene | 34 million | |
| Eocene | 56 million | |
| Paleocene | 66 million | |
| Cretaceous | 146 million | Mesozoic |
| Jurassic | 200 million | |
| Triassic | 252 million | |
| Permian | 299 million | Paleozoic |
| Carboniferous | 359 million | |
| Devonian | 416 million | |
| Silurian | 443 million | |
| Ordovician | 485 million | |
| Cambrian | 542 million | |
| Proterozoic | 2.5 billion | |
| Archean | 4.6 billion | |

Geologists have divided the Earth's long history into smaller, more manageable chunks. Long time spans, such as eons, contain smaller time spans, called eras, which contain smaller slices of time, such as epochs.

Life on Earth first emerged during the Archean eon, which began around 4,000 million years ago. Eukaryotes and complex life forms (see page 16) emerged during the Proterozoic eon, which started 2,500 million years ago.

We are currently in the Holocene epoch, which is a subdivision of the most recent eon. The Holocene spans the last 11,700 years of history, but now some scientists think that this time span has come to an end. They argue that in recent times our planet has changed so much that it's quite different to the Holocene. Humans have caused these changes, so they think this new epoch should be called the **Anthropocene** or 'the age of humans'.

## Geological eras

*Different life forms evolved during different geological periods. They can be seen in the different layers of rock.*

From our humble origins, in the cradle of Africa, modern humans have spread across the globe (see page 121). Earth is now home to 7.7 billion people, and our collective impact is immense.

In recent times, we have levelled forests and cleared huge regions of land to make way for our cities, shopping malls and industrial plants. We have dammed and diverted rivers. We have gouged minerals, metals and other natural resources from the bowels of the Earth.

We have produced enough concrete to cover the Earth's surface in a layer 2 mm (0.08 in) thick. More than 8 billion kg (18 billion lb) of plastic enters our oceans every year, and it's estimated that by 2050, there will be more plastic waste in the ocean than fish.

Our actions are disrupting the global cycling of elements that are needed to sustain life. In the early twentieth century, scientists worked out how to harness nitrogen from the air and turn it into fertilisers and other chemicals. The Haber-Bosch process, as it is called, now removes as much nitrogen from the atmosphere as all of Earth's natural processes combined.

This is causing serious environmental problems. Nitrogen-rich fertilisers run off our fields into the rivers and oceans. This is creating dead zones in the oceans where life is unable to survive. There are more than 400 of these dead zones. The largest, in the Gulf of Oman, is almost the size of Florida and is still growing.

The carbon cycle is in turmoil (see page 135). Thanks to our actions, the concentration of carbon dioxide in the atmosphere is now higher than at any time in the last 800,000 years. This is causing global warming and climate change.

Humans are also directly changing the makeup of life on Earth. Every year, the agricultural sector produces around 4.8 billion farm animals, and 4,800 billion kg (5.3 billion tons) of the world's top five crops: sugarcane, maize, rice, wheat and potatoes. We also pull 80,000 million kg (88.2 million tons) of fish from the oceans, and produce the same amount again through fish farming.

Along the way, we are pushing the world's wild species to **extinction**. Populations of mammals, birds, fish and reptiles have fallen by an average of 60% since 1970, and extinction rates are soaring.

# Climate Change

For millions of years, the carbon cycle was beautifully balanced. The amount of carbon dioxide produced by living things was matched by the amount that was used up. Plants use **carbon dioxide** to power photosynthesis, and vast quantities of carbon dioxide are dissolved in the oceans. Plants and water act as carbon dioxide sinks.

This meant that atmospheric carbon dioxide levels remained more or less the same for a very long time. Now, however, the concentration of atmospheric carbon dioxide is rising.

This is due to human activity. Carbon dioxide is released into the atmosphere when we burn **fossil fuels**, such as coal and oil. At the same time, **deforestation** is a problem. We are cutting down forests so there are fewer plants to absorb this excess.

Since the Industrial Revolution, we have added around 2.2 trillion metric tonnes (2.4 trillion tons) of carbon dioxide into the atmosphere, boosting levels by more than a third. Now the atmospheric concentration of carbon dioxide is higher than at any point in the last 800,000 years. This is leading to **global warming**.

Warming occurs because the carbon dioxide lingers in the atmosphere where it traps some of the Sun's heat. This is called the **greenhouse effect**. Carbon dioxide and methane are both **greenhouse gases**. The greenhouse effect is good, because the warmth it creates makes life on Earth possible, but too much warming is dangerous. **Methane** is released by farm animals and by rice paddies.

Now methane and carbon dioxide are building up in the atmosphere. This is exacerbating the greenhouse effect, and causing global temperatures to rise. The world is now around 1°C (1.8°F) warmer than it was during pre-industrial times.

## The greenhouse effect

*Increasing levels of carbon dioxide in the Earth's atmosphere are causing the world to warm.*

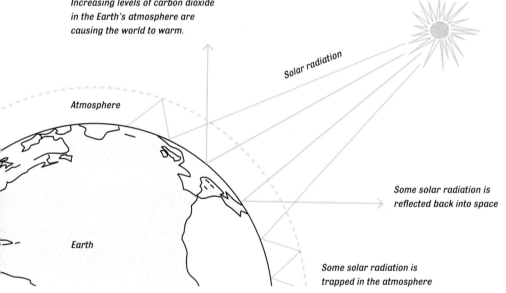

Solar radiation

Atmosphere

Earth

Some solar radiation is reflected back into space

Some solar radiation is trapped in the atmosphere

Twenty of the warmest years on record have occurred in the past 22 years, and the top four years occurred between 2015 and 2018. If this trend continues, scientists estimate that the world could warm up to an extra 5°C (9°F) by 2100.

A temperature rise of a few degrees might not sound like much, but the warming that has already occurred is having a considerable effect.

The climate is changing. We are experiencing more **wild weather** that is less predictable. Many people think that the storms and floods seen globally in the early part of the twenty-first century are evidence of this.

The oceans are absorbing a lot of this excess heat and this is causing the polar ice sheets to melt. Glaciers are melting, too, which means **sea levels rise**. The global sea level has risen by about 20 cm (8 inches) in the last century. Flooding is destroying homes and habitats, and scientists now estimate that the world's seas could rise by more than 2 metres (6.5 feet) by 2100.

## Average global temperatures

*The world has warmed by around 1°C (1.8°F) since pre-industrial times.*

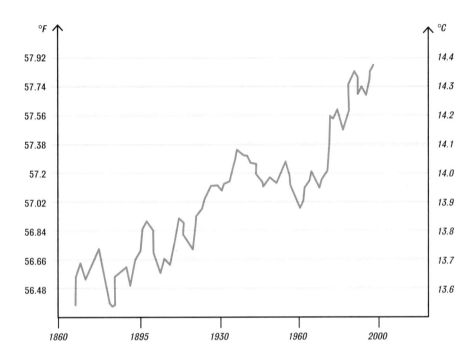

As the oceans absorb carbon dioxide, they are becoming more acidic. This **acidification** of surface ocean waters has increased by about 30% since the start of the Industrial Revolution. This is altering the ocean's chemistry and affecting the life cycles of many marine organisms. On land, animals and plants are struggling to keep up with the pace of climate change.

Climate scientists have suggested that we try to limit global warming to an increase of no more than 1.5°C (2.7°F), but recent reports suggest that this is becoming increasingly unlikely. They think a figure of 3°C (5.4°F) is more likely. Now society needs to decide how it will tackle this problem, and people are talking not of climate change, but of a **climate crisis**.

# Dealing with the climate crisis

Many governments recognise the damage that is being caused by increasing levels of atmospheric greenhouse gases, and are working hard to limit the emission of these gases.

It will take a global commitment to stay below the recommended temperature increase of 1.5°C (2.7°F), but many people believe this is achievable.

## Alarming temperature rises

*Governments are trying to limit global warming to 1.5°C (2.7°F) above pre-industrial levels.*

In 2018, the Intergovernmental Panel on Climate Change (IPCC) published an important report. It outlined what is likely to happen if warming reaches 2°C (3.6°F) or above, and what society can do to prevent this from happening.

At 2°C (3.6°F) of warming, the world will become a profoundly different place. Most of the world's coral reefs will be gone, and the Arctic will become increasingly ice-free during the summer. Huge numbers of animals will become extinct, crop yields will fall dramatically in certain parts of the world, and hundreds of millions of people will be displaced from their homes by rising sea levels.

According to the IPCC report, it will take 'rapid, far-reaching and unprecedented changes in all aspects of society' to prevent this from happening.

Global greenhouse gas emissions need to be dramatically reduced. This will involve moving away from fossil fuels to sources of renewable energy, such as wind power and solar energy. To keep within the 1.5°C (2.7°F) warming limit, it's estimated that **renewable energy** sources will need to supply up to 85% of our electricity needs by 2050. We'll also need to plant many **trees** to help mop up the carbon dioxide in the atmosphere.

By 2050 the planet needs to be at 'net zero'. This means that any carbon dioxide still being pumped into the atmosphere will need to be sucked back out and stored underground. The technology to do this is currently in development.

*Renewable energy sources, such as wind power, can help to combat climate change.*

# Geoengineering

Scientists are looking for additional ways to reduce global warming using technology. This is called geoengineering.

There are two basic approaches. The first seeks to remove carbon dioxide directly from the atmosphere. This could be done by making artificial trees, which are machines that convert carbon dioxide and water into other, more useful products. Another option is fertilising the upper layers of the ocean with nutrients that stimulate the activity of phytoplankton. Phytoplankton pull carbon dioxide out of the atmosphere and use it for photosynthesis. The idea is that when they die, they will sink to the sea floor and take the carbon with them.

The second approach proposes to cool the planet by reflecting more of the Sun's energy back into space. This could be done by placing enormous reflective mirrors on the ground or in space, or by injecting sunlight-reflecting aerosols into the upper atmosphere to mimic the cooling effect of volcanoes.

Space reflectors in orbit

Atmospheric aerosols

Land reflectors

## Future ways to cool the planet

*Sunlight could be reflected into space using mirrors or atmospheric aerosols.*

Both approaches are controversial. Some doubt their effectiveness, and caution they could lead to unforeseen consequences. For example, using sunlight-reflecting aerosols to reduce the amount of sunlight that falls on Earth could slow plant growth, which could be a problem for global food production. It could also change the weather. One study suggested that injecting aerosols into the air could cause profound changes to precipitation, including disrupting the Indian monsoon.

Critics of geoengineering also point out that these strategies address the symptoms rather than the causes of climate change, but advocates point out they could help to buy the world some time while carbon emissions are reigned in. In the meantime, most agree that climate change will not be solved by a single 'silver bullet' and that possible solutions should be carefully explored.

# Extinction and Biodiversity Loss

The passenger pigeon was once the most abundant bird in North America, if not the world. In the nineteenth century, passenger pigeons existed in mind-boggling numbers. Flocks containing hundreds of millions of birds blocked out the Sun and darkened the skies. And then, one day, they were gone. The last ever passenger pigeon, a female called Martha, died at the Cincinnati Zoological Gardens on 1 September 1914.

**Extinction** is commonplace. More than 99% of all the species that have ever lived on Earth are no longer here. New species are evolving all the time, and older ones become extinct when they are outcompeted or unable to adapt to change.

There is a normal, **background rate** of extinction, but sometimes extinction rates spike and vast numbers of species become extinct in relatively short time periods. These events are called **mass extinctions**. There have been five mass extinctions since life on Earth began. The most famous and recent occurred when a massive asteroid slammed into the Earth around 65 million years ago. It caused the demise of the dinosaurs, and three-quarters of all species at the time. The biggest mass extinction event occurred around 250 million years ago, when warming temperatures and one of the biggest volcanic eruptions ever, caused 95% of all species to die.

Now many scientists think we are heading toward a sixth mass extinction that is being caused by our actions. At least 680 vertebrate species have been driven to extinction since the sixteenth century. The passenger pigeon was one of them. Hunters shot, clubbed and trapped the birds, then sold them on as food. For a while, passenger pigeons were the cheapest source of protein in America.

Passenger pigeon

Extinction rates are currently a thousand times higher than during pre-human times, and a million species are now thought to be at risk of extinction, many within the next few decades. This includes around 40% of amphibians, 33% of reef-forming corals, 25% of mammals and 14% of birds.

The picture is less clear for insects, but several studies have documented their decline and conservative estimates suggest that at least one in ten are threatened with extinction.

There are many reasons for this. **Pollution**, climate change, invasive species and **urban development** all play a role, but the biggest drivers of biodiversity loss are agriculture and **overexploitation**.

Overexploitation involves logging, hunting, fishing and gathering species from the wild in a way that is unsustainable. The wild populations that are affected have no time to regrow, reproduce or recover.

**Agriculture** is the next biggest problem. The way we produce our food is having a profound effect on the natural world. Forests are being felled in order to grow crops such as palm oil and soy. Vital carbon sinks, such as forests, are being destroyed, and iconic species such as jaguars, elephants and orangutans are losing their homes.

Today, two-thirds of the world's 70 billion farm animals are reared intensively through **factory farming**, and livestock farming is now responsible

for 14.5% of the world's greenhouse gas emissions. This is more than the global transport sector. Factory farming isn't just harming wildlife. It's making climate change worse.

The health of our ecosystems is deteriorating rapidly. This is a huge problem. The natural world provides us with food, energy, medicines and materials. It gives us clean air, fresh water and soils to grow crops. It regulates the climate, distributes fresh water, and provides us with essential services such as pollination and pest control. Earth's ecosystems are the only natural sinks we have to absorb the carbon dioxide that we jettison into the atmosphere. Nature is vital for our survival.

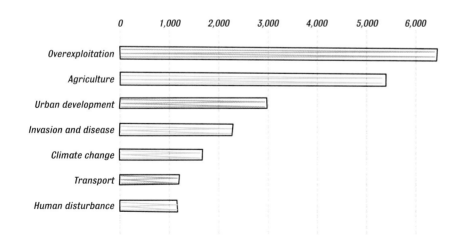

## Activities threatening wildlife

*This chart shows the number of species affected by just seven different activities. Overexploitation and agriculture are the biggest factors threatening biodiversity today.*

# Dealing with biodiversity loss

As it slips through our fingers, people are becoming increasingly aware of the importance of Earth's biological diversity, or biodiversity. The species that we share our planet with are irreplaceable, so people are working hard to stem the tide of biodiversity loss. Modern **conservation** focuses on three different levels: genetic diversity, species diversity and ecosystem diversity.

**Genetic diversity** refers to the amount of genetic variation that is present in groups of living things. When populations dwindle, and species become inbred, genetic diversity is reduced. This puts their future in jeopardy, because it means they may be less able to respond to environmental change.

**Species diversity** refers to the number of species that exist. We talk of species that are 'endangered' or 'threatened'. Endangered species are the ones that are most likely to become extinct, while threatened species are the ones that are likely to become endangered in the near future.

**Ecosystem diversity** refers to all the different ecosystems that exist on our planet, for example, forests, deserts, coral reefs and mangrove swamps. Conservationists consider all of these levels as they monitor ecosystem change and decide which species and ecosystems to support.

Sometimes, approaches that are **low tech**, such as simply protecting key habitats, can be incredibly successful. In 1992, for example, the Canadian government imposed a temporary ban on cod fishing along the country's east coast, after 500 years of fishing had left the cod population seriously depleted. Tens of thousands of people lost their jobs. It was a difficult decision but it paid off. Now stocks are well on the way to recovery. All the fish needed was a little time.

Many endangered species now live in **protected areas**, where they are more or less safe. The world's 1,000 or so remaining mountain gorillas, for example, live in protected national parks in two regions of Africa.

## Mountain gorillas

*The number of mountain gorillas is increasing thanks to conservation.*

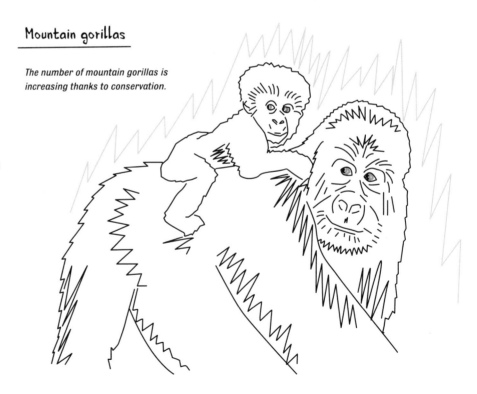

## REWILDING

Rewilding is another low-tech approach to conservation. Rewilding is simply about creating natural spaces where wild species can flourish, and people can reconnect with nature.

Sometimes nature needs a helping hand. If the aim is to create woodland, then trees need to be planted. If the aim is to create paths for migratory fish, then dams may need to be removed. Sometimes, a missing keystone species may need to be reintroduced. Then, over time, as natural processes get going, nature can increasingly be left to look after itself.

Biodiversity grows because natural processes are allowed to establish themselves, and food webs become more complex (see page 132). Now, rewilding is being encouraged as an effective and cost-effective way to increase the range and number of many flagging species.

In the United States, the most famous rewilding example was when wolves were reintroduced to Yellowstone National Park (see page 133). This had a highly positive effect on local levels of biodiversity.

Where Tamworth pigs roam free on a large, rewilded patch of farmland in southern England, they are also boosting biodiversity. The pigs snuffle around in the undergrowth, creating the perfect conditions for fresh sallow saplings to sprout. In turn, these young plants are a food source for the purple emperor butterfly that is now thriving at the site. Now the Knepp Estate, as it is called, is the UK's number one breeding hotspot for this rare insect.

In mainland Europe, an organisation called Rewilding Europe aims to rewild around 10,000 km² (4,000 square miles), of land by 2020. Large, charismatic species are being returned to their former homes. European bison – impressive beasts that once ranged from central Russia to Spain – have been returned to many parts of central and eastern Europe. Beavers have been

## The benefits of rewilding

*When Tamworth pigs were introduced to the UK's Knepp Estate, they created the perfect conditions for purple emperor butterflies to thrive.*

released on more than 150 occasions. Bears now live in more than 20 European countries and wolves have spread across most of the continent.

Rewilding differs from traditional conservation methods, which tend to focus on managed, protected pockets of land. Advocates of rewilding imagine vast networks of land being set aside for nature. This makes rewilding controversial, because some worry that newly reintroduced species could displace local wildlife and come into conflict with farmers.

## GENETIC RESCUE

As science progresses, conservationists are increasingly turning to **high-tech** approaches to help conserve the world's biodiversity. The black-footed ferret is an excellent example. This small, feisty carnivore once inhabited the Great Plains of North America, but in the late 1980s it was driven close to extinction because of disease and because people hunted its main source of prey, the prairie dog.

So, conservationists captured all of the remaining wild animals and used them to establish a captive breeding colony. This was a bold move, because at the time, no one had ever done this before.

Black-footed ferret

They allowed the seven founders to reproduce naturally, but they also used artificial insemination. Semen from key male animals was used to inseminate key females. This was to minimise the effects of inbreeding, and manage the genetic diversity of the emerging population. Now, 30 years later, over 9,000 captive ferrets have been born, and half of them have been released back into the wild.

It has all the makings of a success story, but there's still a problem. Even with artificial insemination, conservationists are still worried that the species lacks genetic diversity, so they're proposing an even more radical solution.

Scientists have cells from two of the original animals that died without reproducing. If they can use the cells for cloning (see page 23), they can create genetic replicas of these animals and effectively boost the size of the founding population from seven to nine. This could make a considerable difference.

In addition, people are discussing whether gene editing can help endangered species (see page 105). The black-footed ferret could have its DNA edited so that it becomes immune to the disease that still threatens it. Similarly, some amphibian species could have their genomes edited to help resist the chytrid fungus that currently blights them. This is called **genetic rescue**.

Prairie dog

# How can we help the planet?

As we enter the Anthropocene epoch, climate change and biodiversity loss are two of the most pressing problems facing our planet. Both rely on fundamental biological processes and both are closely interlinked.

Global change is needed to help redress these problems, but the choices that we make as individuals can also make a difference. Here are five top tips to help you reduce your environmental impact on the planet.

### 1 Eat less meat

Studies suggest that the single, biggest way to reduce your environmental impact is to eat less meat. Eat more plants, and less and better meat from pasture-fed, free-range and organic systems. Buy locally sourced, seasonal produce to reduce the air miles and carbon emissions that are created by transporting food around the globe.

### 2 Waste less

Every year, consumers in rich countries discard over 200,000 million kg (220 million tons) of perfectly edible food. With a little planning and a hearty appetite, this is entirely avoidable.

### 3 Turn the thermostat down

An IPCC report revealed that people tend to overestimate the energy-saving potential of lighting, and underestimate the energy used to heat water. Put on a sweater, and heat your home with energy that comes from renewable sources.

### 4 Change how you travel

Walk and cycle short distances. Use public transport or car share for longer ones. Take trains and buses instead of planes.

### 5 Embrace wildlife

Rewild your garden by sowing local wildflowers and other plants that attract pollinators. Sink a pond; it will attract lots of insects and other invertebrates.

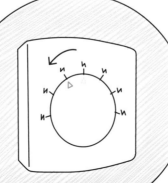

# Glossary

**Active transport** The movement of molecules across a cell membrane from a region of lower concentration to a region of higher concentration. This takes energy.

**Adaptation** The ability of living things to become specialised so they suit their environment.

**Alleles** Different versions of the same gene.

**Amino acid** The chemical building blocks that make up proteins.

**Anabolic pathway** A molecular pathway that builds bigger molecules from smaller ones.

**Angiosperm** A group of seed plants, including grasses, shrubs and most trees, whose seeds develop inside chambers in the flower.

**Anthropocene** The proposed new geological epoch that we are currently living through. A time when humans have become the predominant force shaping the planet.

**Archaea** Primitive, single-celled creatures that are one of the three main divisions of life.

**Arthropod** Spineless animals that have hard outer shells, e.g. insects, spiders and crustaceans.

**Asexual reproduction** The form of reproduction that produces exact copies of the parent; also called cloning.

**Autosome** Any of the chromosomes that are not sex chromosomes.

**Bacteria** Simple, single-celled creatures that are one of the three main divisions of life.

**Biodiversity** All of the life on our planet.

**Biosphere** All of the planet's ecosystems and landscapes added together.

**Cancer** A condition of uncontrollable cell growth.

**Carbohydrate** A simple sugar such as glucose or fructose, which can form the basis of more complex sugars.

**Catabolic pathway** A molecular pathway that breaks big molecules into smaller ones.

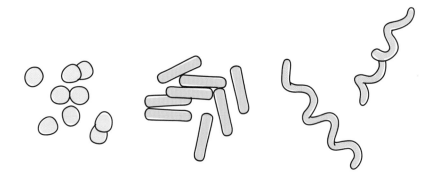

**Central nervous system** The part of the nervous system that consists of the brain and spinal cord.

**Chlorophyll** The green pigment found inside plants; used by plants to make food.

**Chloroplast** Tiny structures found inside plant cells and some algae. The place where photosynthesis occurs.

**Chromosome** A thin strand of DNA that contains genes. In most cells, these are bundled together in pairs in the nucleus.

**Climate change** The recent increase in global temperatures caused by the burning of fossil fuels and other processes; also called global warming.

**Coadaptation** When two separate species develop complementary adaptations because they have evolved alongside one another.

**Communicable disease** A disease, like the common cold, that can be spread from person to person.

**Community** A group that contains populations of different species.

**Complex disorders** Illnesses such as heart disease and stroke, where a mix of genetic and environmental factors influences the disease.

**CRISPR-Cas9** A method to precisely alter the DNA of living things.

**Diffusion** The movement of particles from a more to a less concentrated area.

**Digestive system** A specialised system for digesting food. Everything from the mouth to the anus.

**Diploid** A cell that contains two sets of chromosomes, e.g. skin cells are diploid.

**DNA** Deoxyribonucleic acid is the chemical molecule that encodes life. It is made up of two long molecules, arranged in a spiral.

**Ecology** The study of how living things interact with each other and the environment.

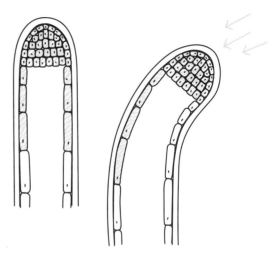

**Ecosystem** A community of living things and their physical environment.

**Endoplasmic reticulum** A network of tiny tubes found inside cells; it is involved in making proteins and lipids.

**Endospore** A tough, dormant structure produced by some bacteria to help them endure harsh conditions.

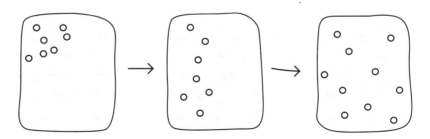

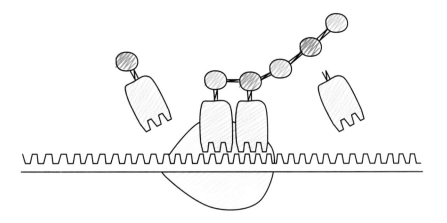

**Gene** A short stretch of DNA that codes for a protein. The basic unit of genetic material that is inherited from our parents.

**Genome** All of the genetic material found inside an organism.

**Geoengineering** Big, bold engineering-based approaches that could possibly be used to help solve climate change.

**Gibberellin** A type of plant hormone involved in growth and development.

**Gravitropism** The ability of plant roots to grow down into the ground.

**Greenhouse effect** The buildup of heat in the atmosphere that is caused by gases such as carbon dioxide and methane.

**Enzyme** A special sort of protein that speeds up chemical reactions.

**Epigenetics** Instructions that change the activity of genes rather than the sequence of DNA.

**Eukaryote** Complex organisms with membrane-bound nuclei, which are one of the three main divisions of life.

**Evolution** The process by which life changes over time and that generates all life on Earth.

**Extinction** When a species dies out it is extinct.

**Fertilisation** The fusion of an egg and sperm to create new life.

**Food chain** A sequence of organisms connected by their feeding habits.

**Fossil fuel** A natural fuel, such as coal or gas, formed in the deep past from the remains of living things.

**Fungus** (*plural* fungi) Eukaryotic life forms such as mushrooms, yeast and moulds.

**Gamete** A specialised sex cell, such as a sperm or egg. It contains half the genetic information of a regular cell.

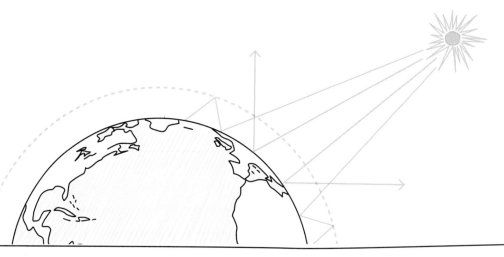

**Greenhouse gas** Gases such as carbon dioxide and methane that are contributing to the warming of the world.

**Gymnosperm** A group of seed plants, such as cycads and conifers, whose seeds are 'naked' and not enclosed in any sort of structure.

**Haploid** A cell that contains a single set of chromosomes, e.g. sperm and egg cells are haploid.

**Heterotrophy** The ability to absorb nutrients directly from the environment, e.g. fungi are heterotrophs.

**Heterozygote** When two different versions or alleles of a gene are present.

**Homeostasis** The ability of living things to maintain a stable, constant internal environment.

**Hominin** Any species of early human that is more closely related to humans than to chimpanzees.

**Homozygote** When two identical versions of the same gene are present.

**Hormone** A chemical messenger that spreads information via the bloodstream.

**Inheritance** The acquisition of certain features from your parents.

**Keystone species** Organisms that have a major effect on their ecosystem. Their disappearance or reintroduction causes widespread change across every level of the ecosystem.

**Lipid** One of the main components found inside living cells.

**Mass extinction** The extinction of a large number of species in a relatively short period of time.

**Meiosis** The form of cell division that leads to the production of sperm and eggs.

**Metabolic rate** The measure of how quickly metabolism occurs.

**Metabolism** All of the life-sustaining chemical reactions that occur inside living things.

**Mitochondrion** (*plural* mitochondria) Tiny, energy-generating structures found inside cells. The place where respiration occurs.

**Mitosis** The form of cell division used by organisms for growth and repair.

**Motor neuron** Nerve cells that carry information from the brain and spinal cord to the rest of the body.

**Multicellular organism** Contains more than one cell.

**Mutation** Any change to the genetic code.

**Mycorrhiza** (*plural* mycorrhizae) An underground fungus that lives symbiotically with plants.

**Natural selection** The process by which organisms that are well adapted to their environment tend to survive and reproduce more than organisms that are less well adapted. This process leads to change over time.

**Nature-nurture debate** The ongoing discussion that attempts to unravel the influence of genetics and the environment on growth and development.

**Neuron** A specialised cell that transmits information to other cells. Also called a nerve cell.

**Non-communicable disease** A disease, such as diabetes, that is not spread between people.

**Nucleotide** The chemical units that make up DNA.

**Nucleus** (*plural* **nuclei**) The cell's control centre; a structure inside the cell that contains DNA.

**Osmosis** The movement of water molecules from a less concentrated solution to a more concentrated solution through a partially permeable membrane; a type of diffusion.

**Pathogen** A disease-causing microorganism such as a bacteria or a virus.

**Peripheral nervous system** All of the nerves that transmit information to the brain and spinal cord.

**Phloem** A type of transport tissue found in plants. It transports food around the plant.

**Photosynthesis** The method that plants use to make food.

**Phototropism** The ability of plants to grow toward the light.

**Population** A group that contains individuals of the same species.

**Prokaryote** Any single-celled organism that lacks a membrane-bound nucleus.

**Protein** One of the main components found inside living cells. Complex molecules made up of chains of amino acids.

**Protist** A diverse group of eukaryotic life forms that are not animals, fungi or plants.

**Reflex** A quick reaction that is not under conscious control.

**Relay neuron** A nerve cell that transmits information between sensory and motor neurons.

**Respiration** The process of obtaining energy from food. Aerobic respiration involves oxygen. Anaerobic respiration does not involve oxygen.

**Rewilding** A form of conservation that involves setting land aside for nature and letting natural, ecological processes take over.

**Ribosomes** Tiny structures found inside cells. The place where proteins are made.

**RNA** Ribonucleic acid is similar to DNA. A messenger molecule that carries instructions from DNA to control the production of proteins.

**Sensory neuron** A nerve cell that transmits information from the sense organs to the brain and spinal cord.

**Sex chromosome** The chromosomes that determine the sex of an organism.

**Sexual reproduction** The form of reproduction that leads to offspring with genetic information from both parents.

**Species** A group of living things with similar characteristics. They can breed together to produce more living things of the same type.

**Stem cell** A cell that can divide to form different specialised sorts of cell and copies of itself.

**Stomata** Tiny holes found in plant cells that allow gases to enter and leave the plant.

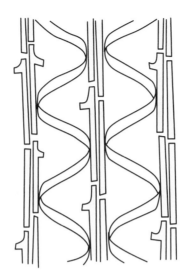

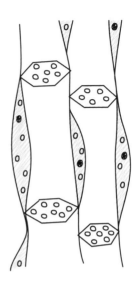

**Symbiosis** A relationship between two or more organisms that live closely with one another.

**Taxonomy** The study of how life is interrelated and classified.

**Transcription** The process that converts DNA to RNA inside cells.

**Transitional fossil** A fossil that shows intermediate features from two different groups of animals; evidence of evolution in action.

**Translation** The process that converts RNA to protein inside cells.

**Variation** Any difference between cells, organisms or groups of organisms that is caused by genetic or environmental factors. It is a key part of evolution.

**Vertebrate** Any animal with a backbone.

**Vestigial feature** A feature that has no current purpose, but that used to serve a purpose in the evolutionary past.

**Xylem** A type of transport tissue found in plants. It moves water from the roots to the leaves.

**Zygote** The cell that is formed when gametes fuse during sexual reproduction.

# Index

commensal
  relationships 15
common ancestors 109,
  110, 116, 117, 120
communicable diseases
  80, 82–3, 84–8
communities 122, 124, 125
competition
  122, 126, 130–1
competitive inhibition 58
complex disorders 95, 104
compound eyes 75
cones 66, 75
confocal microscopes 28
conformers 53, 65
connective tissue 39, 42–3
conservation 136, 146–7
consumers 122, 132
corals 20, 125, 142, 145
corneas 66, 74, 75
creationism 108, 112, 113
CRISPR-Cas9 95, 105
crustaceans 11, 20
cyanobacteria 15
cystic fibrosis 81, 89, 100
cytoplasm 28, 31, 33, 34,
  37, 59, 60, 99
cytoskeleton 29

differentiation 40
diffusion 24, 36, 41, 45
digestion 20, 27, 44
digestive system
  38, 41, 44–5
direct disease
  transmission 80, 84
disease 80–93
  prevention
    80, 85, 92–3, 97
  risk factors
    81, 83, 89, 92–3
  transmission 80, 84
DNA 25, 27, 28, 33,
  59, 94, 96–9
  cell division 34–5
  damage 81, 89, 91
  epigenetics 107
  evolution 116, 117, 120
  fossils 120
  junk 95, 98
  twin studies 106
  *see also* genetics;
    mutations
Dolly the sheep 23
dominant alleles
  95, 100–1
double circulatory system
  38, 46–7, 65
Down syndrome 104
drugs 59, 80

electron microscopes 28
electron transport
  chain 60
embryos 23, 34, 35,
  40, 98, 105, 117
endocrine glands 76–7
endocrine system
  67, 69, 76–7
endoplasmic reticulum
  25, 28, 30
endospores 14
endothermic reactions
  52, 62
energy
  activation 56, 57
  metabolism 52, 53
  renewable 137,
    142, 149
  respiration 61
environment, genetics
  94, 104, 106, 107
enzymes 26, 27, 63
  digestion 44–5
  genes 97
  homeostasis 69
  inhibitors 53, 58–9
  metabolism
    52, 54, 55, 56–9
epidermal cells
  48, 50, 51
epigenetics 7, 95, 107
epithelial tissue 39, 42
ethanol 53, 61
ethene 66, 79
eukaryotes 11, 13, 16–17,
  20, 27–30, 34, 139
evolution 12, 20, 22,
  55, 59, 108–21, 128–9
  in action 108, 115,
    118–19
  competition 130
  humans 20, 109, 120–1
evolutionary trees
  108, 111, 116

exothermic reactions
  53, 60
extinction 7, 108, 114,
  136, 139, 144–5
eyelight fish 15
eyes 66, 74–5, 117

expression 40, 94,
  98, 107
gene therapy 105
genetic rescue
  137, 148
genetics 7, 94–107
  diversity 136, 146, 148
genomes 94, 96, 97, 107
genus 12, 13
geoengineering
  137, 143
gibberellins 66, 79
glia 39, 43
global warming
  137, 140–1, 142–3
glucose 55
  control 68, 77
  metabolism 52, 53
  photosynthesis 62–3
  respiration 55, 60–1
glycogen 26
glycolysis 55, 60
Golgi bodies 25, 28, 30
gonorrhoea 80, 86
gravitropism 66, 79
greenhouse effect
  137, 140
greenhouse gases
  137, 140, 141, 142, 145
growth 12, 24, 40
  cell division 34
  plants 48–9, 79
gymnosperms 11, 19

heterotrophs 17, 18, 20
heterozygotes 95, 101
hibernation 53, 65
HIV/AIDS 21, 80, 84, 87
Holocene 139
homeostasis 66, 68–9,
    78–9
hominins 109, 120, 121
*Homo neanderthalensis*
    12, 109, 120–1
*Homo sapiens* 12, 20,
    109, 120
homozygotes 95, 101
hormones 66, 76–7
    cancer 91
    plants 78–9
horse evolution 115
hoverflies 129
Huntington disease 103
hygiene 80, 85, 86
hyphae 18, 23
hypothalamus 67, 69, 72

I

immune system
    15, 46, 87, 90, 120
indirect disease
    transmission 80, 84
influenza virus 21, 84
ingestion 12, 20
inheritance 100–1,
    109, 110
inherited diseases
    81, 89, 95, 100, 103
inhibitors of enzymes
    53, 58–9
insects 11, 12, 20, 21, 145
insulin 27, 77
interdependence 122, 125
invertebrates 10, 20
IPCC report 142, 149
iris 66, 74

J

Japanese knotweed 126

K

kakapos 131
keystone species 122, 133
kingdoms 13, 16
Krebs cycle 60

L

lactic acid 53, 61
Lamarckism 108, 112
large intestine 38, 45
leaves 30, 37, 49, 50–1,
    62, 79, 127, 129, 131
lens 66, 75
life expectancy 82
lifestyle 81, 83, 89, 97, 104
light microscopes 28
lignin 32
Linnaeus, Carl 11, 13
lipids 26
Lister, Joseph 85
liver 45, 53, 61
lock and key model
    52, 56–7
long-sightedness 75
Lucy 109, 120
lungs 38, 46–7
lysosomes 29

M

macromolecules 26
malaria 80, 82, 88
malignant tumours 81, 90
mammals 10, 20,
    65, 139, 145
    adipose cells 26
    clones 23
    forelimbs 117
mass extinctions 136, 144
measles 80, 82, 87
medulla 67, 72
meiosis 22, 24, 35, 102

melatonin 76–7
Mendel, Gregor
    95, 100, 101
meristems 48
mesophyll 39, 48, 51
messenger RNA (mRNA)
    27, 99
metabolic rate 53, 64–5
metabolism 52–65, 68
metastasis 81, 90
methane 137, 140
microbes 12, 53, 61
microscopes 28
mimicry 122, 129
minerals 123, 127
mitochondria 16, 24, 28,
    29, 30, 31, 34, 43, 59, 60
mitosis 22, 24, 34, 40, 48
mixotrophs 17
molluscs 10, 20
mosquitoes 80, 88
motor neurons 67, 70, 71
mouth 38, 44
movement 12
MRI 67, 73
MRSA 80, 86
multicellular life 16, 20, 41
multi-enzyme complex 59
muscles
    cells 25, 29, 40
    eye 75
    fatigue 61
    shivering 69
    system 39, 41
    tissue 42, 43
mushrooms 23
mutations 95, 103, 104,
    109, 116, 117, 118, 121
mutualistic relationships
    15, 18
mycelium 18
mycorrhizae 11, 18
myelin sheath 30
myxozoa 20

N

natural selection
    109, 111, 113, 116,
    118, 120, 121, 124
nature-nurture debate
    94, 106, 107
Neanderthals
    12, 109, 120–1
negative feedback
    loops 69
nerve cells *see* neurons
nervous system
    39, 41, 69, 70–3, 76
nervous tissue 42, 43
neurons 20, 25, 29, 30, 39,
    40, 43, 67, 70, 71, 72
neurotransmitters 71
nitrification 134
nitrogen 15, 129, 139
nitrogen cycle 123, 134
non-communicable
    diseases 81, 82–3,
    89–93
non-competitive
    inhibition 58
non-vascular plants
    11, 19
nucleic acids 26, 27
nucleotides 96, 103
nucleus 16, 25, 28, 30

O

obesity 81, 92–3
oesophagus 38, 45
oestrogen 76, 77
omnivores 20
*On the Origin of Species*
    109, 111, 113
optic nerves 66, 74
optogenetics 67, 73
organ systems 38, 41
organelles 29, 30, 62
osmosis 24, 36–7
ovaries 23, 35, 67, 77

overexploitation
    136, 145
oxygen 36, 41, 46–7, 50
    metabolism 52, 53
    photosynthesis 62–3
    respiration 60–1

P

palisade layer 51
pancreas 66, 77
pandas 55
parasites 15, 18, 20
passenger pigeons 144
Pasteur, Louis 85
pasteurisation 85
pathogens 80, 83, 84, 85,
    87, 123, 126
peppered moths 108, 118
peptidoglycan 33
peripheral nervous system
    (PNS) 70
pH 53, 57, 68, 123, 127
phages 21
phloem 25, 31, 32, 48, 51
phospholipids 26, 28
photosynthesis
    17, 19, 30, 31, 32, 49,
    50, 51, 54, 59, 62–3
    carbon cycle 135
    cyanobacteria 15
    metabolism 52
phototropism 66, 78
physical activity 81, 89, **93**
physiological adaptations
    123, 128
phytoplankton 16, 143
pigs, Tamworth 147
pineal glands 67, 76
pitcher plants 129
pituitary glands 67, 72, 76
plants 11, 12, 16, 19,
    25, 126
    active transport 37
    cells 30–1, 33
    climate change 140

## Acknowlededgments

Thanks to Kate Duffy and Lindsey Johns for their brilliant editorial and artistic skills. And thanks to my family and my dog. For being there.